IEC NATIONAL

2008/2009

Apprenticeship Curriculum

Student Manual

Year 4

[Cat. No. A400-08/09]

Copyright © 2008 by the Independent Electrical Contractors, Inc. All rights reserved.

IEC

NATIONAL

IEC Apprenticeship Curriculum Conditions of Use

1. The publication date for the 2008/2009 edition of IEC's apprenticeship guide is June 2008. The materials included therein are believed to be current and reflective of current regulatory requirements and applicable codes. Users should contact Independent Electrical Contractors, Inc., Attention: Apprenticeship and Training Department, 4401 Ford Avenue, Suite 1100, Alexandria, Virginia 22302 to ascertain that these materials are still current prior to use. These materials are sold with the understanding and agreement of the purchaser that IEC does not warrant or represent that they are free from errors or conflicts. Users are encouraged to report any errors, conflicts, or omissions to IEC, Attention: Apprenticeship and Training Department, so that errors can be corrected.

2. The IEC Apprenticeship Program is designed with a number of specific components; each of which must be used if the apprentice is to gain full benefit from the curriculum. Specifically, latest editions of both the student and instructor curriculum guides must be used to ensure that all key components are covered. While lessons may be rearranged within a curriculum year, and requirements of a local nature may be incorporated into the curriculum as required, all material within a curriculum year must be covered within that school year to ensure that the apprentice is properly prepared to progress into the subsequent year's program.

3. Current editions of the curriculum guides must be purchased and implemented annually to help ensure that students are being taught with the most up-to-date materials. Students must be provided or required to obtain the textbooks cited in the curriculum. Teaching from out-of-date editions of either the curriculum or the textbooks may result in students being taught outdated subject matter that might conflict with current practices. The electrical industry is continually changing. New materials and processes are being introduced. Standards are being revised, new standards are being introduced, and local, state and federal requirements are subject to continual revision. Textbooks are updated on an irregular basis.

4. Students should be encouraged to retain all of their texts for reference throughout their apprenticeship and working career, but should be advised to check current codes, regulations, etc. for specific questions or problems.

5. While the apprenticeship training program with its curriculum guides and related texts incorporates a significant degree of safety training, it does not, nor is it intended to, provide all of the types of safety training required for contractor employees under OSHA, EPA, and similar regulations. Contractors and employees are encouraged to refer to applicable local, state, and federal regulations to ensure that they are in compliance with all applicable safety and health guidelines.

6. Upon evidence that a chapter, program sponsors, educational entity, or other customer is not abiding by these conditions, or is no longer in good financial standing with IEC, IEC reserves the right to refuse to provide future curriculum, texts or manpower in support of that individual's/organization's apprenticeship program.

7. Copyright 2008. Independent Electrical Contractors, Inc. ("IEC"), 4401 Ford Avenue, Suite 1100, Alexandria, Virginia 22302, phone: (703) 549-7351, fax: (703) 549-7448. All rights to this work are reserved. No part of this work may be reproduced or copied in any form or by any means—graphic, electronic or mechanical, including photocopying, recording, taping, or use on an information retrieval system—without written permission obtained in advance from IEC.

Approved by IEC National Apprenticeship and Training Committee—January 19, 2004
Approved by IEC National Board of Directors—May 1, 2004

WE NEED YOUR FEEDBACK!

YOU MAY USE THIS FORM TO REPORT ANY QUESTIONS, COMMENTS OR CORRECTIONS
YOU HAVE REGARDING THE IEC APPRENTICESHIP CURRICULUM

EVERY IDEA OR QUESTION WILL BE CONSIDERED
(We even want to know about typographical errors, misspelled words, etc.)

CURRICULUM YEAR: _____ Edition 2008/2009 _____

TEXTBOOK NAME _____ TEXTBOOK PAGE NUMBER(S) _____

Type of error is:

☐ Typo   ☐ Content   ☐ Test Question   ☐ Other

DEAR CURRICULUM COMMITTEE: _____
_____
_____
_____
_____
_____
_____
_____
_____
_____
_____

Signed: _____ Chapter: _____

Mail to: IEC, Inc., 4401 Ford Avenue, Suite 1100
                        Alexandria, VA 22302
Fax to: (703) 549-7448
Email comments to: Trainingeditor@ieci.org

Copyright © 2008 by the Independent Electrical Contractors, Inc. All rights reserved.

IEC NATIONAL PRIDE

Electrical Curriculum
Year Four
Student Manual

INTRODUCTION – STUDENT MANUAL

Congratulations. You are almost there. As a fourth-year apprentice, you have succeeded in completing three years toward becoming an electrician and will soon be ready to take your journeyman's exam. You are taking on the responsibility to complete a final training course of 144 hours per year of classroom time and as many hours of study and homework. But you are past the worst part. Much new material was presented in the first three years. The fourth year will involve some review of this material, some new material and lots of practice at solving problems and interpreting code. Even more than the first three years, you should have increasing opportunities to use the information from class to do troubleshooting on the job.

The training materials used in this curriculum assume that you have completed the first three years of training and remember the information that was taught during these years. You may want to take some time to quickly review the previous curricula in order to refresh your memory.

In this fourth year of the IEC electrical apprenticeship, you will learn about control devices and power quality. You will also have the opportunity to develop your leadership skills and to review for the Journeyman's Examination. Throughout the fourth year, you should be working at a higher level of skill than in previous years. You should be analyzing (breaking down), synthesizing (putting back together), and evaluating (troubleshooting) information in each lesson.

For example, you should be able to discuss circuit design as it applies to a particular installation. You should then be able to draw the circuit. Finally, you should be able to wire the circuit and troubleshoot the circuit if it does not work according to specifications. You should be able to know what the components are and how they work together. In other words, you should be able to apply both theory and code to the installation of electrical circuits and equipment in various settings.

Each lesson tells you the purpose, objectives, content, and homework. There are also sections for professionalism, relevant tools, *Code*® references, and study tips. Additionally, each lesson contains a "Tool Box Talk" that concerns safety, tricks of the trade, and plans, specifications, and documentation. You will be expected to fully participate in class activities and discussions. Questions are always encouraged. Since you will be doing more problem-solving and trouble shooting, don't hesitate to ask your instructor for additional help in solving problems that you encounter on the job or in doing the homework. You should know that the best students are often the students that ask the most questions.

You can use the back of each sheet of paper for taking notes. Additional notes can be kept on additional pages that you add to your notebook. In this way, all of your study materials are in one place, making it easy for you to review the information and to study for tests and quizzes. If you have done this kind of note-taking during the previous three years, you will find it easy to review the material that you need to remember for the fourth year and your journeyman's exam.

Tests and quizzes will be given throughout the course to help you and your instructor determine how well you are learning and remembering the material in the course. You will be expected to maintain a 70 percent average on all the tests. In addition to midterm and final semester tests, quizzes may be given without prior notice or review. For this reason, it is very important that you keep up-to-date on your homework.

This course is designed to produce top quality electricians who will become the supervisors of the future. Your employer has cared enough to provide you with this opportunity. Now it is your responsibility to make the most of this opportunity. This introduction is designed to help you know how to make the best use of this curriculum. Please review it carefully.

The following is a detailed preview about each of the sections of this curriculum.

Purpose

The purpose of each lesson is a brief summary of the instructional intent for the lesson.

Objectives

The objectives are designed to tell you what abilities and/or competencies you should have by the end of each lesson. Test questions for this course are specifically designed to test your abilities as stated in the objectives. Therefore, it is worth spending some time making sure that you understand and can do each of the things stated in the objectives. These will form the basis for your test questions and, ultimately, your grade in the class.

As you progress through each year, these objectives are becoming increasingly more difficult. This is done to match your growing skills as an electrician. Correspondingly, test questions in the later lessons will be harder than questions in the earlier lessons. This year, you will be required to do more problem solving than you have done in the previous two years

Content

Knowledge and Skill

This is the content that you will be learning in this lesson. By the end of the lesson, you should feel that you understand all the items listed here. This section expands on the skills listed in the objectives.

Introduction – Student Manual – Page 3

This section should also help give you an overview of the whole lesson. This sections emphasizes both skills and theory.

It would be very helpful to use this section as the outline for your notes for this lesson.

When reviewing your lessons, this section will give you a quick overview of the material that is covered in the lesson.

Professionalism

This section refers to the attitudes or work traits that you will need to develop when you use the knowledge and skills listed above.

These traits are the ones you are most likely to need while you are doing work on the job.

The "Professionalism Box" will provide training in how to better communicate, what it means to be a supervisor, goal setting, and other details of being a journeyman electrician. In order to grow to be a competent journeyman you must increase your people skills and your productivity skills. These actions will establish you "on the road" to successful Foremanship.

These lessons are intended as an introduction only. There is much to learn about each of these subjects. Management is a complex function and requires years of study and practice to master. Hopefully, these sections will provide a good starting point for further study.

Relevant Tools and Equipment

This is a list of the equipment or tools that you will need to complete this lesson.

Make sure that you bring these items to class unless you know that the instructor has these items for you to use in class.

Copyright © 2008 by the Independent Electrical Contractors, Inc. All rights reserved.

☞ Code References

These are the parts of the National Electrical *Code®* that pertain to this lesson. Be sure to review these sections before going to class. Know how to use and interpret code is a very important part of the lessons for fourth year. It is an even more important part of the journeyman's exam.

As you know, the *NEC®* is one of the most basic tools of an electrician. It is very important that you know the *NEC®* and how to use the handbook.

Many test questions come directly from the *NEC®*, so this section can also help you review for tests and quizzes.

Don't forget the index in the back of the *Code®* book. Learn to use this index. Your ability to use this index will have a big impact on how well you do with many of the lessons in the fourth year.

It is highly recommended that you use the *Code®* book in conjunction with all of your textbooks. This will greatly enhance your learning experience. You will also learn how to use and interpret the *NEC®* on the job.

Homework

📖 1) Reading Assignment

Read the following section of the text books or other referenced materials:

This section details the reading assignments to be completed for homework. Completion of homework assignments is part of your grade. You should complete the homework before it is covered in class. This will help you understand and remember ideas that are covered in class.

AA 2) Key Terms

These terms are the new terms that you will need to know in order to understand this lesson. Only some of them will be found in the *Illustrated Dictionary for Electrical Workers*. The rest will be found in your other textbooks. Use the index or glossary in each textbook to help you define each word.

Writing the definition for each of these terms before doing your homework will help you to remember and use these terms. Before doing your homework, make sure that you know the definition for each term. You don't have to use only the terms listed in the lesson. If you find other terms that you haven't had before, you may want to add them to your list.

Copyright © 2008 by the Independent Electrical Contractors, Inc. All rights reserved.

Introduction – Student Manual – Page 5

Some of the words listed in these lessons have already been used in years One, Two and Three. You may want to refer to your class notes for help with these terms. Research shows that the more terms that you know and can use correctly in the class or on the job, the more likely you are to be successful. In other words, if you want to be an electrician, you have to talk like an electrician.

3) Practice Exercises

Answer the following referenced questions in the texts or other materials:

This section details the questions from the textbooks that are to be completed and brought to class the next week. Your instructor will collect these exercises as part of your grade.

Study Tips

Included with the homework are study tips. These are designed to make studying easier and to help you remember what you study. You may find some tips to be more useful than others. The key is to try them more than once. Sometimes it takes practice to make them work. If after several tries, the techniques still don't help you, move on and try other study tips until you find methods for studying that truly help you study faster, easier and with better results.

Tool Box Talk

Safety

Job site safety is the most important function of each worker. To facilitate job site safety practices a safety topic, called a "Tool Box Talk" will be discussed in class with each lesson. Read these safety topics closely, and pay attention in class when they are being discussed. They could save you a lot of pain; toes, fingers, eyes, or limbs; a lot of time off work; or even your life - or the life of your co-worker. Take safety to heart - although accidents do happen, they don't have to happen and can be avoided!

Tricks of the Trade

This section attempts to provide you with some operating practices that could make you more professional or productive. Some are very elementary. Some you may already know, or have already tried. Others may be new to your experience. All are worth considering.

Copyright © 2008 by the Independent Electrical Contractors, Inc. All rights reserved.

Plans, Specifications, and Documentation

This section asks specific questions regarding the use of construction documents and blueprints. Sometimes they refer to the *Code*®. Others times they refer to the various blueprints that were used in 3rd Year as teaching aids. Try to answer these questions, in an informal manner (unless assigned as homework by your instructor) and ask questions in class the next week if you have problems answering the questions or understanding the concept presented.

IEC NATIONAL

FOURTH YEAR LESSONS

LESSON	CONTENT	CLASS DATE	INSTRUCTOR SIGNATURE
401	Orientation and Safety		
402	Community First Aid & Safety		
403	Community First Aid & Safety		
404	Solid State Electronic Control Devices		
405	Electromechanical and Solid State Relays		
406	Advanced Controls Lab #1		
407	Photoelectric and Proximity Controls		
408	Programmable Controllers		
409	Advanced Controls Lab #2		
410	Mid-Term Review and Exam		
411	Reduced Voltage Starting		
412	Accelerating and Decelerating Methods		
413	Advanced Controls Lab #3		
414	Preventative Maintenance and Troubleshooting		
415	Advanced Controls Lab #4		
416	Leadership		
417	Semester Review		
418	Semester Exam		
419	Using Digital Multimeters to Diagnose Power Quality		
420	*National Electrical Code®* and Related Standards, Safety Regulation, Power Systems		
421	Power Distribution Systems		
422	Services, Switchboards, and Panelboards		
423	Conductors and Overcurrent Protection Devices		
424	Grounding		
425	Designing and Installing Wiring Systems		

Copyright © 2008 by the Independent Electrical Contractors, Inc. All rights reserved.

LESSON	CONTENT	CLASS DATE	INSTRUCTOR SIGNATURE
426	Branch and Feeder Circuits		
427	Mid-Term Review and Exam		
428	Receptacle and Lighting and Switching Outlets		
429	Motors and Compressor Motors		
430	Hazardous Locations		
431	Hazardous Locations, Special Types		
432	Signs and Sign Connections		
433	Load Calculations		
434	Final Code Review and Test Preparation		
435	Fourth Year Review		
436	Fourth Year Final Exam		

Copyright © 2008 by the Independent Electrical Contractors, Inc. All rights reserved.

Bibliography

TEXTBOOKS

Listed below are the textbooks referred to in the lessons for Year Four:

Community First Aid and Safety, American Red Cross

Community CPR, American Red Cross

Illustrated Dictionary for Electrical Workers, 2nd Edition. Patricia A. Titus, James E. Titus, John Traister, Albany, NY; © 2002

National Electrical Code® 2008; National Fire Protection Association, Quincy, MA; © 2007

Stallcup's Electrical Design Book; James Stallcup; National Fire Protection Association: Quincy, Massachusetts, © 2008

Electrical Motor Controls for Integrated Systems, 3rd Edition; Glen Mazur and Gary Rockis; American Technical Publishers, Inc., Homewood, IL; © 2005 (Both text and workbook)

FOR INSTRUCTORS

National Electrical Code® Handbook 2008; National Fire Protection Association, Quincy, MA.

Solid State Fundamentals For Electricians; Gary Rockis; American Technical Publishers, Inc., Homewood, IL © 2001

Practical Guide to Electrical Grounding; ERICO Publication, Solon, OH; © 1999

Power Quality Measurment and Troubleshooting; 2nd Edition; Glen Mazur; American Technical Publishers, Inc., Homewood, IL ©1999

* The certified American Red Cross Instructor will supply American Red Cross books.

IEC Electrical Curriculum
Year Four Student Manual

Lesson 401 – Orientation and Safety

Purpose

This lesson will give a brief overview of both instructor and student obligations to the apprenticeship training program. The lesson will also review the advantages of being part of the Independent Electrical Contractors (IEC) apprenticeship program. Students will also go over their responsibilities regarding safety and handling of hazardous material on the job site.

Objectives

By the end of this lesson, you should be able to:

401-1 Summarize policies and procedures of the IEC chapter
401-2 Review the advantages of being part of the Independent Electrical Contractors (IEC) apprenticeship program
401-3 Review using a Hazcom manual on the job
401-4 Maintain safe practices on the job and deal appropriately with unsafe conditions that may arise

Content

Knowledge and Skills

- Understanding what is expected of you as a fourth year IEC apprentice and student.
  ➢ Procedures
  ➢ Classroom rules

- What you can expect of the instructor

- Remind you why you have chosen the right trade—independently.

- What are the advantages of being part of the Independent Electrical Contractors (IEC).

- Review of OSHA regulations regarding the handling of hazardous materials on the job.
  - IEC Hazcom Manual
  - MSDS
- Reminders of what makes for a safe environment on the job site.
  - Personal responsibility
  - Safety Practices

Professionalism

- Ability to speak effectively with others.

Speaking Effectively with Others

Communication is an important part of your job, including working with your team and with customers. Here are some keys to talking so others will listen.

- **Determine why you are speaking.** What is the purpose? What you talking about? Why is it important?

- **Know your customer or audience.** Who will listen to your message? What is important to this person or group? How will the audience feel about your message?

- **Prepare your message.** Gather information and plan your key points. Be brief, be clear, and have the facts to back up what you say. After you tell someone something, pause. If you don't pause, they can't give you feedback or ask questions. Practice pausing for three to four seconds and see how much more often people interact with you.

- **Check your spoken image and your body image.** Work to eliminate bad habits such as rocking, saying "uh" or looking down while speaking or listening. Your body language can speak as "loudly" as the words you say.

Practice, Practice, practice. As a professional, you may have to deliver news to your customer that they don't want to hear! Being prepared is key to doing this well.

Relevant Tools and Equipment

- None

Code References

- None

Orientation and Safety (Student Manual) Lesson 401-3

Homework

1) Reading Assignment

- None

2) Key Terms

Define the following key terms:

- Hazcom
- OSHA
- MSDS
- IEC
- Independent

3) Practice Exercises

Answer the following referenced questions in the texts or other materials:

- None

Study Tips

▶ Be sure to read your homework assignment before class. By reading the homework assignment, you will know which concepts you don't understand and need to ask questions about. Make a list of the things you don't understand so that you can ask your instructor about them before or during class. Good students learn to ask questions of their instructors. It is how they make sure that they understand everything before they have to take a test.

▶ You will be learning calculations for single phase transformers in this lesson. Spend some time before class with these formulas so that you will able to use them with ease during class time.

▶ You will learn a lot of new concepts in this lesson. There are memory tips that can help you to remember all these terms.

- If you want to remember something, review it immediately, and then again within 20 minutes.
- Use a spaced practice schedule—20 minutes a day for five days is much better than two hours of memorization at one sitting.
- Make the information meaningful. Don't try to memorize information that you don't understand.

Copyright © 2008 by the Independent Electrical Contractors, Inc. All rights reserved.

- Organize the information into chunks of seven items or less. If you have to memorize a big list of something, group into smaller lists organized in some logical fashion even if only you know that logic.

Tool Box Talk

Safety

Remember that different areas in the electrical trade have different safety requirements and practices. Sometimes these requirements and practices are generally known, such as requirements given by *NEC®* or OSHA. Sometimes the company that manufactures materials or products will have special safety procedures and may even require special training before you use them. Finally, there are many ways to stay safe that you can learn from experienced tradesmen in the on-the-job (OJT) portion of your apprenticeship. The key to safety is to always stay alert, use common sense, and be sure to ask questions whenever you are unsure about a product or a material.

Tricks of the Trade

Skilled electricians have the ability to troubleshoot and do problem solving across all technical areas. This apprenticeship program will give you the opportunity to build your knowledge and skills in all these areas. In addition to technical and troubleshooting skills, successful electricians have the ability to work with people—both the people they work with and customers. This apprenticeship program will provide you with opportunities to work as a member of a team, build your communication skills, and develop positive relationships with others. Take advantage of every opportunity you have to build your "people skills" both in the classroom and on the job.

Plans, Specifications, and Documentation

As an apprentice you are learning a number of technical competencies that build from a basic understanding of electricity and mechanical and structural knowledge. This will include the ability to read plans, specifications, and forms of documentation used in the electrical industry. You will also need to be able to read, interpret, and sometimes make changes on blueprints and plans used by other trades. Be aware of technical areas that may involve the ability to read plans, specifications, and documentation, especially those areas that involve other trades.

Community First Aid and Safety (Student Manual) Lesson 402-1

| **I E C** NATIONAL | Electrical Curriculum Year Four Student Manual |

Lesson 402 – Community First Aid and Safety

Purpose

To have students receive American Red Cross certification in Community First Aid and Safety. (Note: A certified American Red Cross Instructor should teach this lesson and Lesson 403.)

Objectives

By the end of this lesson, you should be able to:

402-1   Recognize emergency situations
402-2   Be prepared for emergencies
402-3   Know the steps for taking action: Check, Call, and Care
402-4   Know how to deal with conscious and unconscious victims
402-5   Know how to clear an airway
402-6   Know how and when to perform rescue breathing
402-7   Know how and when to perform CPR

Content

Knowledge and Skills

- Recognizing emergencies and being prepared to act.
- Emergency action steps.
- How to check the victim.
- How to respond to adult life-threatening emergencies.
- How to respond to heart problems.
- Children with life-threatening emergencies.
- How to care for children with breathing emergencies.
- How to care for infants with breathing emergencies.

Copyright © 2008 by the Independent Electrical Contractors, Inc. All rights reserved.

Community First Aid and Safety (Student Manual) Lesson 402-2

Professionalism

- Learn the important steps in project planning
- Learn how to develop and implement project planning tasks

Project Management

The professionalism lessons in the next several lessons will give you an introduction to some of the important aspects of project management. The purpose of these mini-lessons is to help you if and/or when you are asked to help manage a project at work. In addition to these mini-lessons, you should pay attention to how project management is done at your company. For instance, how does your company chart the progress of a project? Paying attention to these things now will help you when you are asked to step up to the plate and become a project supervisor.

Relevant Tools and Equipment

- Red Cross Mannequins and related supplies

Code References

- None

Homework

Reading Assignment

Read the following section of the text books or referenced materials:

- Read *Community First Aid and Safety* as assigned by your instructor

Community First Aid and Safety (Student Manual) Lesson 402-3

Key Terms

Define the following key terms:

- AED
- Abdominal thrusts
- Chest compression
- EMS
- Heart attack
- HIV
- Stoma
- AIDS
- CPR
- Good Samaritan Laws
- Heimlich maneuver
- Rescue
- Breathing

Practice Exercises

Answer the following referenced questions in the texts or other materials:

- None

Study Tips

▶ It is important that you ask questions if you don't understand something. The certified American Red Cross Instructor is trained to answer questions and help you understand the material. Take advantage of this training to be sure you know how to perform these important first aid and CPR techniques.

Tool Box Talk

Safety

✗ The risk of disease transmission during CPR training is extremely low according to the Manual Guidelines 2000 for Cardiopulmonary Resuscitation and Emergency Cardiovascular Care published in the American Heart Association's journal *Circulation* in August 22, 2000. In addition, the use of CPR manikins has never been shown to be responsible for an outbreak of infection, and a literature search through March 2000 revealed no reports of infection associated with CPR training.

Tricks of the Trade

✓ The American Red Cross minimizes the risk of disease transmission during CPR training through the following process:

- The American Red Cross develops and delivers courses and trains instructors to provide for the safety of all participants. This includes minimizing the risk of disease transmission. The use of manikins has never been documented as being responsible for transmitting a case of bacterial, fungal or viral disease.

- The American Red Cross develops standards for decontaminating manikins based on information from the Centers for Disease Control and Prevention. These standards are consistent with the current American Heart Association Emergency Cardiovascular Care Manual Guidelines. Red Cross instructors are given specific manikin cleaning procedures to follow before, during and after class.

- American Red Cross chapters and trained instructors follow rigorous manikin cleaning procedures.

Plans, Specifications, and Documentation

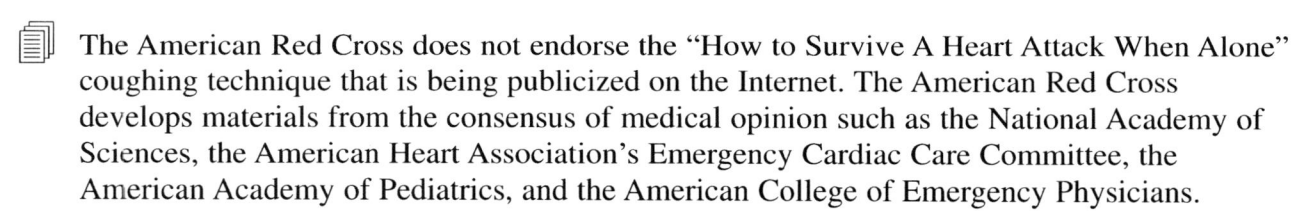

The American Red Cross does not endorse the "How to Survive A Heart Attack When Alone" coughing technique that is being publicized on the Internet. The American Red Cross develops materials from the consensus of medical opinion such as the National Academy of Sciences, the American Heart Association's Emergency Cardiac Care Committee, the American Academy of Pediatrics, and the American College of Emergency Physicians.

The 1992 Guidelines for Cardiopulmonary Resuscitation and Emergency Cardiac Care and the Guidelines 2000 for Cardiopulmonary Resuscitation and Emergency Cardiovascular Care International Consensus on Science briefly discuss the technique called Cough CPR. Cough CPR is a self administered form of cardiopulmonary resuscitation described by CM Criley in 1976. According to Criley, self initiated CPR is possible; however, its use is limited to clinical situations in which the patient has a cardiac monitor, the arrest is recognized before loss of consciousness, and the patient can cough forcefully. To date, there is insufficient scientific research concerning the efficacy of Cough CPR. Therefore, the American Red Cross does not advocate teaching the technique until it has been thoroughly tested in national studies and found to be effective.

Community First Aid and Safety (Student Manual) Lesson 403-1

| **I E C** **PRIDE** **NATIONAL** | Electrical Curriculum

Year Four
Student Manual |

Lesson 403 – Community First Aid and Safety

Purpose

To have students receive American Red Cross Certification in Community First Aid and Safety. (Note: A certified American Red Cross Instructor should teach this lesson and Lesson 402.)

Objectives

By the end of this lesson, you should be able to:

403-1 Identify the causes, and how to reduce the risk of injury
403-2 Know about, and how to care for, cuts, scrapes and bruises
403-3 Know about the types and causes, and how to care for, burns
403-4 Know about, and how to care for, muscle, bone, and joint injuries
403-5 Know how to recognize and care for sudden illness
403-6 Know how to prevent poisoning; and/or treat it, if it occurs
403-7 Know about heat and cold related illnesses and how they should be treated
403-8 Know how to recognize and care for the special problems of the young and elderly

Content

Knowledge and Skills

This lesson is designed to instruct you in the following:

- The causes and how to reduce the risk of injuries.
- Cuts, scrapes and bruises and how to care for them.
- Burns and how to care for them.
- Injuries to muscles, bones and joints.
- Recognizing and caring for sudden illness.
- Poisoning prevention and care.
- Heat and cold related illness.
- Recognizing and caring for special problems of the young and elderly.

Copyright © 2008 by the Independent Electrical Contractors, Inc. All rights reserved.

Professionalism

- Ability to plan before beginning work
- Ability to identify and use proper equipment and tools

Project Planning

Planning is an important part of any project. Have you ever experienced situations like this? "If only I had thought to bring this tool, the job would be a cinch!" Or "Why didn't I check the amount of cable 1 needed before I left for the work site? Now I need to make two trips!" Then you know how important planning can be. Planning can have a big influence on your work. This is why companies spend a great deal of time and effort on project management.

Project planning includes all activities that result in a course of action. Goals for the project must be set and their priorities established. Goals include resources to be committed, completion times, and activities. Areas of responsibility must be identified and assigned. The work done on a project each day has an effect on the overall project plan. How does your daily work affect "the big picture?"

Relevant Tools and Equipment

- Bandages
- Splints
- Blankets
- Pillows
- First Aid Kit

Code References

- None

Community First Aid and Safety (Student Manual) Lesson 403-3

Homework

1) Reading Assignment

- Read *Community First Aid and Safety* as assigned by your instructor.

2) Key Terms

Define the following key terms:

- Absorption
- Avulsion
- Capillary
- Fracture
- Hypoglycemia
- Incision
- Ingestion
- Injection
- Seizure
- Sprain
- Vein
- Artery
- Bruise
- Dislocation
- Hyperglycemia
- Hypothermia
- Infection
- Inhalation
- Laceration
- Splint
- Strain

3) Practice Exercises

- None.

Study Tips

▶ Be sure to pay attention to the pictures and graphics in the First Aid texts. These are almost as important as the text. They have valuable information to which you should pay attention.

Tool Box Talk

Safety

✗ The American Red Cross encourages the public to recognize the signals of a heart attack. You should remember these too. Someday, they might save a life.

- Persistent chest pain or discomfort (which can range from discomfort to an unbearable crushing sensation in the chest) that lasts longer than 3 to 5 minutes or is not relieved by resting, changing position, oral medication, or goes away and then comes back.

- Discomfort, pain or pressure in either arm; discomfort, pain or pressure that spreads to the shoulder, arm, neck or jaw.

- Breathing difficulty, which may cause dizziness.

- Nausea.

- Skin appearance, which may be pale or bluish in color. The face may be moist or may sweat profusely.

- Unconsciousness.

Tricks of the Trade

✓ To care for a heart attack victim, take the following steps.

- Recognize the signals of a heart attack.
- Call 911 or the local emergency number for help.
- Convince and help the victim to stop activity and rest comfortably.
- Try to obtain additional information about the victim's condition.
- Assist with medication, if prescribed.
- Monitor the victim's condition.
- Be prepared to give CPR and use an AED if the victim's heart stops beating.

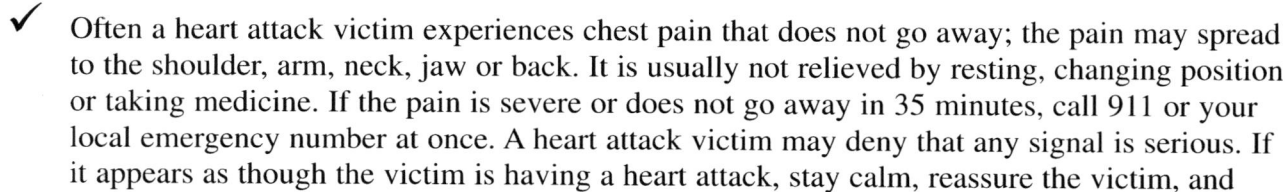

✓ Often a heart attack victim experiences chest pain that does not go away; the pain may spread to the shoulder, arm, neck, jaw or back. It is usually not relieved by resting, changing position or taking medicine. If the pain is severe or does not go away in 35 minutes, call 911 or your local emergency number at once. A heart attack victim may deny that any signal is serious. If it appears as though the victim is having a heart attack, stay calm, reassure the victim, and call 911 or your local emergency number!

Plans, Specifications, and Documentation

There are several OSHA Standards that are concerned with safety and first aid. Here are a few that are very important for electricians.

- Medical services and first aid—OSHA 1910.151

- Electric Power Generation, Transmission, and Distribution—OSHA 1910.269 (b)

- First aid and medical attention—OSHA 1926.23

- First Aid Kits—OSHA 1926.50

Solid State Electronic Control Devices (Student Manual) Lesson 404-1

IEC NATIONAL

Electrical Curriculum

**Year Four
Student Manual**

Lesson 404 – Solid State Electronic Control Devices

Purpose

To teach students the different types of solid state control devices, and their connection and operation in a circuit.

Objectives

By the end of this lesson, you should be able to:

404-1 Describe the basic printed circuit (PC) board and its main components
404-2 Analyze semiconductor theory and its relation to semiconductor devices
404-3 Categorize N-type and P-type material
404-4 Classify rectification systems
404-5 Differentiate among various types of diodes
404-6 Explain the theory, operation, and use of various other solid state devices, including TRIACs, DIACs, and transistors

Content

Knowledge and Skills

This lesson is designed to instruct you in the following:

- What are some of the normal components of a printed circuit (PC) board?
  - Insulated board
  - Foils or traces
  - Pads
  - Terminal contact
  - Edge Cards

Solid State Electronic Control Devices (Student Manual) Lesson 404-2

- What are the components of a semiconductor and how do they work with each other to form a semiconductor?
  - Atom
  - Protons
  - Neutrons
  - Electrons
  - Valence electrons

- What are the types of polarity of semiconductor materials?
  - Coding
  - N-type material
  - P-type material

- What is a rectifier and what types of rectifier systems are used?
  - Definition
  - Half-wave rectifier
  - Full-wave rectifier
  - Three-phase rectifier

- What are some of the different types of diodes and how do they work?
  - Diode
  - Zener diode
  - Light-emitting diode
  - Photoconductive diode

- What are other types of solid state components and how do they work?
  - Transistors
  - Amplifiers
  - SCRs
  - TRIACs and DIACs
  - I.C.s and OP-amps
  - Logic gates
  - Various light-activated devices

- Triacs, Diacs and their operating characteristics?
  - Definition
  - Construction
  - Operation
  - Applications
  - DIAC testing

- What is a unijunction transistor (UJT) and what are its operating characteristics and applications?
  - Unijunction transistor definition
  - UJT biasing
  - Theory of operation
  - UJT application

Solid State Electronic Control Devices (Student Manual) Lesson 404-3

- How can a transistor be used as an AC amplifier?
  - Amplifier gain
  - Bandwidth
  - Decibel
  - Transistor amplifiers
  - Common base amplifiers
  - Common collector amplifiers
  - Class of operation
  - Input and output impedances
  - Transistor specifications
  - Transistor testers and service tips

Professionalism

- Ability to plan before beginning work
- Ability to identify and use proper equipment and tools
- Willingness to organize work to make it as functional as possible
- Ability to communicate with installation team, including project managers

Gannt Charts

The Gantt chart is a model of project planning named for its creator, Henry Gantt. The Gantt Chart is a bar chart that depicts the relationship of activities over a time period.

Many companies use project planning or scheduling charts that you will be expected to read and use as part of your job. One of these is a Gantt chart. The Gantt chart uses graphics to display the duration of activities. This kind of Gantt chart might look like the one below:

New office:	101
	01 02 03 04 05 06 07 08 09 10 11 12
Department Layouts and plan	▬
Identify equipment to order	▬
Order equipment and modular furniture	▬
Install electrical services	▬▬
Install telecommunication services	▬▬
Install computer network	▬▬
Test systems	▬▬

Copyright © 2008 by the Independent Electrical Contractors, Inc. All rights reserved.

Solid State Electronic Control Devices (Student Manual) Lesson 404-4

Relevant Tools and Equipment

- Hand tools
- Soldering iron and solder
- Alligator clip leads
- Digital or analog Volt-Ohm meter
- Oscilloscope, if available
- Wire jumpers

Code References

- None

Homework

1) Reading Assignment

- Read *Electrical Motor Controls for Integrated Systems*, Chapter 12. Do all the questions at the end of the chapter.

2) Key Terms

Define the following key terms using either the *Illustrated Dictionary for Electrical Workers,* the Glossary in the Appendix or your textbooks:

- Anode
- Cascaded amplifiers
- Cut off region
- Diodes
- Fiber optics
- Gain
- Holding current
- Laser
- Opto coupler
- Pads
- Photoconductive cell
- Integrated circuits
- LED
- PC boards
- Photovoltaic cell
- Phototriac
- Saturation region
- Rectifiers
- Semiconductor
- Thermistor
- Transistor
- Unijunction transistor (UJT)
- Amplifier
- Cathode
- DIAC
- Distortion
- Foils
- Gates
- Silicon-controlled rectifier
- Traces
- TRIAC
- Valence electrons
- Zener diodes

Copyright © 2008 by the Independent Electrical Contractors, Inc. All rights reserved.

Solid State Electronic Control Devices (Student Manual) Lesson 404-5

3) Practice Exercises

Answer the following questions:

- Complete TECH-CHEK 12 and worksheets 12-1 through 12-5 in *Workbook for Electrical Motor Controls for Integrated Systems*.

Study Tips

▶ You are using two books for the first time in this lesson. Take some time to get to know these books. How are the chapters constructed—are there chapter purpose statements? How important are the graphics and illustrations in the chapters? Are there review questions? Is there a Glossary? What is included in the Appendix? How could you use the information in the appendix? How is the Workbook set up? Is there a difference between the Worksheets and the Tech-Cheks? Spend some time thinking about how you can best use the information in the books and how you can best study using these books

Tool Box Talk

Safety

✗ It is very important for you to get immediate treatment for every injury, regardless how small you may think it is. Many cases have been reported where a small unimportant injury, such as a splinter wound or a puncture wound, quickly led to an infection, threatening the health and limb of the employee. Even the smallest scratch is large enough for dangerous germs to enter, and in large bruises or deep cuts, germs come in by the millions. Immediate examination and treatment is necessary for every injury.

✗ As with getting medical attention for all injuries, it is equally important that you report all injuries to your supervisor. It is critical that the employer check into the causes of every job related injury, regardless how minor, to find out exactly how it happened. There may be unsafe procedures or unsafe equipment that should be corrected.

Copyright © 2008 by the Independent Electrical Contractors, Inc. All rights reserved.

Tricks of the Trade

✓ Equipment damage is more often caused by differences in voltage than by surges in current. Transient voltages (ground potential rise), which appear throughout the ground path itself, cause more damaged equipment, lockups and garbled data than any other phenomenon. Current flowing in power circuits, grounds from in-rush, short circuits and lightning cause transient voltages. Understanding that grounding and bonding are used to equalize the potential differences in the grounding system is important.

Plans, Specifications, and Documentation

OSHA Standard 1910.333 (a) (1) specifies that live parts of an electrical installation should be de-energized before anyone works on them, unless it can be demonstrated that de-energizing introduces additional or increased hazards or is infeasible due to equipment design or operational limitations. Live parts that operate at less than 50 volts to ground need not be de-energized if there will be no increased exposure to electrical burns or to explosion due to electric arcs.

Electromechanical and Solid State Relays (Student Manual) Lesson 405-1

IEC NATIONAL

Electrical Curriculum

Year Four
Student Manual

Lesson 405 – Electromechanical and Solid State Relays

Purpose

To teach the student about electromechanical and solid state relays and using the proper application of these relays.

Objectives

By the end of this lesson, you should be able to:

405-1   Organize the operations and functions of various electromechanical relays
405-2   Utilize proper contact arrangement and terminology
405-3   Describe the operation and functions of solid state relays
405-4   Demonstrate the selection and installation of the proper relay for an application
405-5   Identify the advantages and disadvantages of different types of relays

Content

Knowledge and Skills

This lesson is designed to instruct you in the following:

- What is an electromechnical relay and what are the different types?
  - Definition
  - Reed
  - General-purpose
  - Machine Control
- What is meant by contact arrangement?
  - Poles
  - Throws
  - Breaks

- What is a solid state relay and what are its capabilities?
  - Definition
  - Comparison
  - Input signal
  - Response
  - Voltage and current rating
  - Voltage drop
  - Insulation and leakage
  - Zero current turn-off
  - Zero voltage turn-on
  - Solid state control
  - Solid state relay design
  - Engineering specifications
  - Troubleshooting
- How do you choose the right relay for the application?
  - Environment
  - Response time
  - Loading
  - Numerous other factors
- What are the advantages and disadvantages of electromechanical and solid state relays?
- Trouble shooting

Professionalism

- Ability to plan before beginning work
- Ability to identify and use proper equipment and tools
- Demonstrate safe working habits
- Willingness to organize work to make it as functional as possible
- Ability to communicate with installation team, including project managers

Gantt Charts

The Gantt chart shown in Lesson 404 is the simplest form of Gantt chart. The light colored solid bars show the completed part of the task, black bars show overdue parts of the task and the bars with stripes show uncompleted parts of the task.

New office:	101
	01 02 03 04 05 06 07 08 09 10 11 12
Department Layouts and plan	
Identify equipment to order	
Order equipment and modular furniture	
Install electrical services	
Install telecommunication services	
Install computer network	
Test systems	

Relevant Tools and Equipment

- Continuity meters
- Continuity beepers

Code References

- None

Homework

1) Reading Assignment

- Read *Electrical Motor Controls for Integrated Systems*, Chapter 14, pp. 375-408.

Electromechanical and Solid State Relays (Student Manual) Lesson 405-4

2) Key Terms

Define the following key terms:

- Break
- Electromechanical relay
- Heat sink
- Hybrid solid state relay
- Optical isolation
- Pole
- Reed relay
- Response time
- Snubber circuit
- Solid state relay
- Suppression
- Throw
- Transient voltage
- Voltage drop dissipation
- Zero voltage turn-on
- Zero current turn-off

3) Practice Exercises

Answer the following questions:

- Tech-Chek 14 and worksheets 14-1 through 14-5 in *Workbook for Electrical Motor Controls for Integrated Systems* book.

Study Tips

▶ Review the questions at the end of each chapter before you read the text. This will help you to focus on the key topics in the textbook, to give you key terms to look for, and to take notes on the information.

▶ Use the questions at the end of the chapters as your preview and your outline for the reading. The answers to the odd numbered review questions are included at the back of the book. Check your own answers against these answers. Use this as a way to confirm that you understand the information in the textbook. This will also tell you if you are on the right track for answering the homework questions.

Tool Box Talk

Safety

✗ Bleeding is the most visible result of an injury. Each of us has between five and six quarts of blood in our body. Most people can lose a small amount of blood with no problem, but if a quart or more is quickly lost, it could lead to shock and/or death. According to the American Red Cross, one of the best ways to treat bleeding is to place a clean cloth on the wound and apply pressure with the palm of your hand until the bleeding stops. You should also elevate the wound above the victim's heart, if possible, to slow down the bleeding at the wound site. Once the bleeding stops, do not try to remove the cloth that is against the open wound as it could disturb the blood clotting and restart the bleeding. For more information, see www.redcross.org.

Tricks of the Trade

✓ Be aware of your surroundings and that conditions can change quickly as voltage and current levels vary in individual circuits. Never assume a circuit is de-energized or that it is operating at the correct levels of voltage and current. An installer should beware of taking measurements in humid or damp environments and ensure that no atmospheric hazards such as flammable dust or vapor are present in the area.

Plans, Specifications, and Documentation

When fixed or electrical equipment is de-energized for work, the circuits energizing them must be locked out or tagged out or both. This includes stored electrical energy from sources such as capacitors, which should be discharged before work begins. A lock and a tag should be applied on each disconnecting means used to de-energize circuits and equipment on which work is to performed. A qualified person must test the circuits and equipment to make sure the circuits have been properly de-energized. Likewise, before energizing, a qualified person should conduct tests and visual inspections to make sure that all tools, electrical jumpers, shorts, grounds, and other such devices have been removed, so that the circuits and equipment can be safely energized.

Advanced Controls/Lab #1 (Student Manual) Lesson 406-1

| *I E C* NATIONAL | Electrical Curriculum Year Four Student Manual |

Lesson 406 – Advanced Controls/Lab #1

Purpose

This lesson is intended to allow the student to use lab facilities to practice, experiment, and apply the knowledge learned in the previous two lessons. Specifically, the students will experiment with safely working with solid state control devices and making connections in circuits of solid state and electromechanical relays.

Objectives

By the end of this lesson, you should be able to:

406-1   Utilize solid state control devices
406-2   Operate electromechanical and solid state relays

Content

Knowledge and Skills

This lesson is designed to instruct you in the following:

- Experiment, hook up and practice using solid state control devices.
- To allow the student to utilize solid state control devices.
- Experiment, hook up and understand the operation of electromechanical and solid state relays.

Professionalism

- Willingness to organize work to make it as functional as possible
- Ability to communicate with installation team, including project managers

Strengths and Weaknesses of the Gantt Chart

- The strength of Gantt charts is that they are able to show the status of each project at any point in time and that they can show overlapping or parallel tasks.
- The weakness of Gantt charts is that they don't do a very good job of showing how activities relate to each other except in terms of time. It is also difficult to incorporate unforeseen activities. Gantt charts show timelines for individual activities but don't really show whether the whole project is on time, behind time or ahead of schedule.

Relevant Tools and Equipment

- Basic hand tools
- Multimeter
- Special lab tools, as needed

Code References

- None

Homework

1) Reading Assignment

- None

2) Key Terms

Review the terms in Lessons 404 and 405.

3) Practice Exercises

Perform lab work based upon Chapters 11 and 12 in EMC.

Advanced Controls/Lab #1 (Student Manual) Lesson 406-3

Study Tips

- Don't be afraid to use your textbook as an active study tool.
- Underline important points.
- Put question mark next to material that you don't understand.
- Circle key words and phrases
- Refer to the Glossary when you come across a word that you don't understand or a word that is included in the Key Terms section of your homework.
- Put a star in the margin to emphasize certain graphics or illustrations.

Tool Box Talk

Safety

✗ If an accident occurs with bleeding that is very serious, the American Red Cross suggests applying pressure to the nearest major pressure point, located either on the inside of the upper arm between the shoulder and elbow, or in the groin area where the leg joins the body. Direct pressure is better than a pressure point or a tourniquet because direct pressure stops blood circulation only at the wound. Only use the pressure points if elevation and direct pressure haven't controlled the bleeding. Never use a tourniquet (a device, such as a bandage twisted tight with a stick, to control the flow of blood) except in response to an extreme emergency, such as a severed arm or leg. Tourniquets can damage nerves and blood vessels and can cause the victim to lose an arm or leg. For more information, see www.redcross.org.

Tricks of the Trade

✓ Scaffolding may be set up and taken down many times during a project. Because scaffolding is considered temporary equipment and is moved from place to place at the job site, it is no surprise that sufficient care is not always taken in erecting it. As a result, inadequate scaffolding is responsible for many construction site accidents. Be sure to take extra precautions when using scaffolding on the job site and comply with all OSHA requirements for the use of scaffolding (OSHA 1926 Subpart L).

Plans, Specifications, and Documentation

If work is to be performed on overhead lines, the lines should be de-energized and grounded before the work is started. OSHA Standard 1910.333 also specifies distances that should be kept between a worker and overhead lines. The distance may vary according to voltage but a good rule of thumb is to keep at least five feet between yourself and overhead lines when working on the ground and ten feet if you are working in or with mechanical equipment, such as an aerial work platform.

Photoelectric and Proximity Controls (Student Manual) Lesson 407-1

| *I E C* *PRIDE* *NATIONAL* | **Electrical Curriculum** **Year Four** **Student Manual** |

Lesson 407 – Photoelectric and Proximity Controls

Purpose

To introduce the students to photoelectric and proximity controls, and how to install and connect them in an electrical circuit.

Objectives

By the end of this lesson, you should be able to:

407-1   Evaluate when and how to use a photoelectric control in a control circuit
407-2   Explain how the different types of photoelectric controls function
407-3   Discriminate among the different types of proximity switches and how they function
407-4   Explain the Hall effect

Content

Knowledge and Skills

This lesson is designed to instruct you in the following:

- What is a photoelectric control and what are the different types of sensors?
    - Definition and operation
    - Photovoltaic
    - Photoemissive
    - Polarized
    - Convergent
    - Specular
    - Diffused

Copyright © 2008 by the Independent Electrical Contractors, Inc. All rights reserved.

- What are the different photosensor scanning techniques and light sources?
  - Direct
  - Reflective
  - Retroreflective
  - Polarized
  - Convergent
  - Specular
  - Diffused

- What are the different types of proximity switches and how do they work?
  - Inductive
  - Capacitive
  - Hall effect

- What is the Hall effect?
  - Theory
  - Types of Hall effect sensors
  - Application
  - Installation
  - Troubleshooting

- Photoelectric and proximity outputs

- Troubleshooting

Photoelectric and Proximity Controls (Student Manual) Lesson 407-3

Professionalism

- Ability to plan before beginning work
- Willingness to organize work to make it as functional as possible
- Ability to communicate with installation team, including project managers

PERT Charts

Setting Priorities: The PERT charts that you will learn about in this lesson represent plans of action that may make a very specific impact on the crew or team that you work with. This is why it is important for you to know how to read this kind of chart—as well as Gantt charts and flow diagrams. The reason for developing these charts is to help set priorities and schedule actions for work.

You can also use PERT charts to set priorities for your own work. You need to think about and set out your priorities by determining what is most important. Then you can think about how to spend your time during the day to accomplish the most important things in an effective and efficient manner. You may point out that you don't always determine your tasks for the day. Someone else may be assigning these to you. This doesn't mean that you can't plan actions and set priorities within your own work responsibilities.

In his book, *Getting Things Done When You Are Not In Charge*, Geoffrey Bellman talks about things that you can do to "perform to priorities." He suggests that you focus on your unique contribution. Managers and supervisors track what is important to them. They may be tracking your work. They may be following what you produce and considering your impact on projects. Think about how what you do is important to your company.

Take time in your work to think about how you do things: how much time does it take for you to do a certain task? How much are you costing the company in terms of time and supplies? How could you do things better or quicker? How can you make certain that you get things done when they need to be done?

If you learn to link your work to key management systems, you will be a more valuable and more employable worker. And maybe, one day, you will be the one using tools like PERT charts to make management decisions.

Relevant Tools and Equipment

- Basic hand tools

Photoelectric and Proximity Controls (Student Manual) Lesson 407-4

Code References

- None

Homework

1) Reading Assignment

- Read *Electrical Motor Controls for Integrated Systems*, Chapter 15, pp. 411-443.

2) Key Terms

Define the following key terms:

- Actuation
- Capacitive proximity sensor
- Convergent beam scan
- Current sinking output
- Current sourcing output
- Dark-operated
- Dielectric
- Diffuse scan
- Direct scan
- ECKO
- Hall effect
- Inductive proximity sensor
- Interfacing
- Interference
- Light-operated
- Modulated
- Pendulum actuation
- Photovoltaic
- Photosensor
- Photoelectric
- Polarized scan
- Proximity switch
- Reflective scan
- Retroreflective scan
- Scanning
- Specular scan
- Unmodulated
- Vane Actuation

3) Practice Exercises

Answer the following questions:

- Complete Tech-Chek 15 and the worksheets for Chapter 15 in *Workbook for Electrical Motor Controls for Integrated Systems*.

Photoelectric and Proximity Controls (Student Manual) Lesson 407-5

Study Tips

▶ In the exercises for this chapter, you are required to design and draw circuits from the information on a data sheet. If you need a review on drawing circuit diagrams, review Chapter 4 and 5 of *Electrical Motor Controls for Integrated Systems*.

Tool Box Talk

Safety

✗ Shock can threaten the life of the victim of an injury if it is not treated quickly. Even if the injury doesn't directly cause death, the victim can go into shock and die. Shock occurs when the body's important functions are threatened by not getting enough blood or when the major organs and tissues don't receive enough oxygen. Some of the symptoms of shock are a pale or bluish skin color that is cold to the touch, vomiting, dull and sunken eyes, and unusual thirst. Shock requires medical treatment to be reversed, so all you can do is prevent it from getting worse.

✗ The American Red Cross suggests several steps for controlling shock. You can maintain an open airway for breathing, control any obvious bleeding and elevate the legs about 12 inches unless an injury makes it impossible. You can also prevent the loss of body heat by covering the victim (over and under) with blankets. Don't give the victim anything to eat or drink because this may cause vomiting. Generally, keep the victim lying flat on the back. A victim who is unconscious or bleeding from the mouth should lie on one side so breathing is easier. Stay with the victim until medical help arrives. For more information, see www.redcross.org.

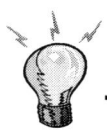

Tricks of the Trade

✓ To prevent accidents associated with scaffolding, a number of safety precautions should be taken during its construction and use.

1. To avoid the use of makeshift platforms, plan each job carefully to ensure that scaffolding is used only when required, and that the scaffolding conforms to all applicable construction and safety regulations.
2. Scaffolds should be designed, built and inspected by trained and experienced workers.
3. Place base plates, sills or footers on solid ground; make sure the scaffold is leveled or plumbed.
4. Do not use damaged end frames or braces. Make sure to attach braces at all points provided; do not shortchange bracing.
5. Tie the scaffold to the structure being worked on, if possible.
6. Use only scaffold-grade wood or metal catwalks for platforms. Before each job, inspect each scaffold platform thoroughly for breaks, knots, cracks or warped boards.
7. Planks should have cleats attached permanently to keep them from sliding off the scaffold frame.
8. Use guardrails and toe boards on platforms higher than 6 feet.

Plans, Specifications, and Documentation

OSHA Standard 1910.333 requires illumination of spaces that contain exposed electrical parts. Also, anyone working in a confined or enclosed space, such as a manhole or vault, should use protective shields, barriers, or insulating materials to protect inadvertent contact with exposed energized parts. Doors, hinged panels, and similar features should be secured to prevent them swinging into a person and causing contact with exposed energized parts.

t last, it's easier to save electrical energy.
nd yours.

lighting CONTROL

ducing Siemens i-3 Control Technology™: simple, cost-effective
ting control to reduce energy costs. Easy set up via network,
interface, and integrated touch panel means you can have a
ens i-3 system up and running in minutes. Easy installation
s on labor. And modular design based on off-the-shelf
ponents saves on maintenance.

more information or to contact a sales representative,
e-mail or visit our web site.

SIEMENS

ens Energy & Automation Inc. • 1-800-964-4114, Ref. Code: I3IEC • info@sea.siemens.com • www.sea.siemens.com/i-3

©2008 Siemens Energy & Automation, Inc.

Programmable Controllers (Student Manual) Lesson 408-1

IEC NATIONAL PRIDE

Electrical Curriculum

Year Four
Student Manual

Lesson 408 – Programmable Controllers

Purpose

This lesson will give you an introduction to programmable controller components and operation.

Objectives

By the end of this lesson, you should be able to:

408-1   Explain the uses of the programmable controller
408-2   Relate the functions of the parts of a programmable controller
408-3   Identify I/Os and write a simple program for a programmable controller
408-4   Categorize programmable controller applications
408-5   Identify the advantages of using multiplexing for specific applications

Content

Knowledge and Skills

This lesson is designed to instruct you in the following:

- What is a programmable controller and where is it used?
  - Definition
  - History
  - Process manufacturing
  - Discrete parts manufacturing
  - Microcomputers
- What are the main parts and functions of a programmable controller?
  - Power Supply
  - Input-output interface
  - Processor section
  - Programming section
  - Inputs and outputs

- How do you program a programmable controller?
  - Basic logic functions
  - Basic programming
  - Manufacturer's programming
  - Memory

- Where are programmable controllers used?
  - Manufacturing
  - Industry
  - Fluid power control
  - Security
  - Home use

- What is multiplexing? Where and how is it used?
  - Definition
  - Uses
  - Wiring

- Troubleshooting

Professionalism

- Ability to plan before beginning work
- Willingness to organize work to make it as functional as possible
- Ability to communicate with installation team, including project managers

Programmable Controllers (Student Manual) Lesson 408-3

PERT Charts

Many companies use project planning or scheduling charts that you will be expected to read and use as part of your job. One of these is a PERT Chart. PERT is an acronym that stands for "Program Evaluation and Review Technique." It was begun by the Polaris missile program in the mid-1950s. When multiple projects have to take place, involving multiple steps and people, the object of the chart is to bring it all under control.

A PERT chart involves four elements: Boxes or circles in which activities are written, Lines showing the direction of progress, Dates indicating completion targets or deadlines and, optionally, names of people who are responsible for each activity.

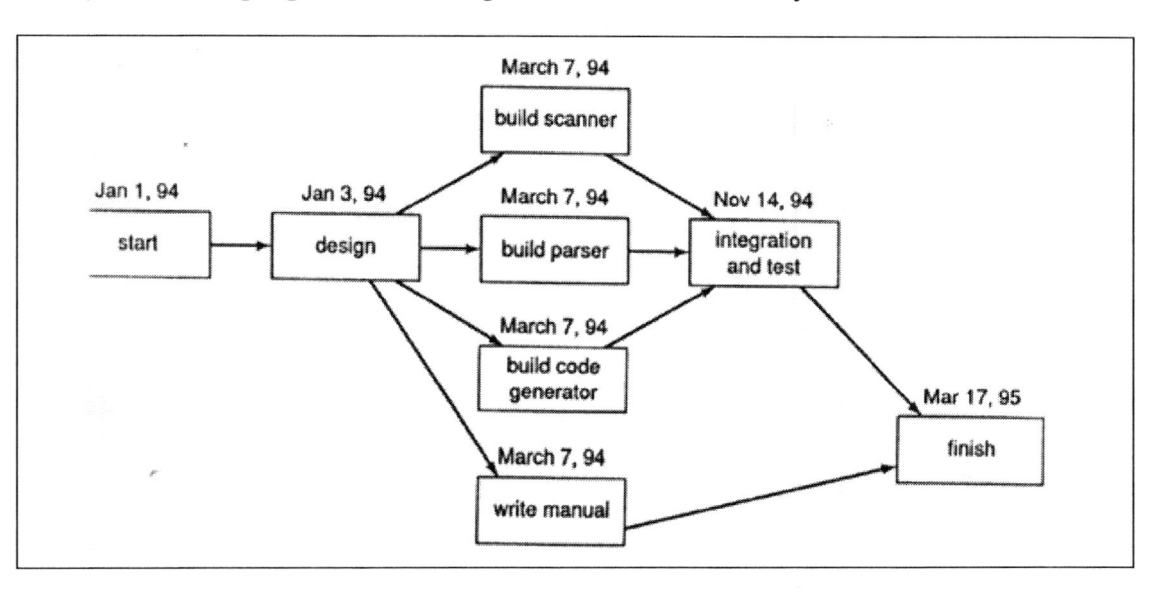

Relevant Tools and Equipment

- None

Code References

- None

Homework

1) Reading Assignment

- Read *Electrical Motor Controls for Integrated Systems*, Chapter 15, pp. 445-483.

2) Key Terms

Using either the *Illustrated Dictionary for Electrical Workers*, the Glossary in the Appendix or your textbooks, write definitions for each of these terms before doing your homework:

- Analog
- Data I/O
- Discrete I/O
- Electrical noise
- Input/output status indicator
- Multiplexing
- Processor section
- Programmable controller
- Scan

3) Practice Exercises

Answer the following questions:

- Complete Tech-Chek 16 and the worksheets for Chapter 16 in *Workbook for Electrical Motors Controls for Integrated Systems*.

Study Tips

▶ To do the exercises for the Workbook, you are asked to write a number of programs. It will help you do this more easily if you review all of Chapter 16 before you attempt to do the homework.

Programmable Controllers (Student Manual) Lesson 408-5

Tool Box Talk

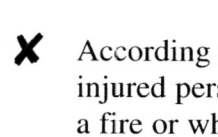

Safety

✘ According to the American Red Cross, in the event of an accident, you should move the injured person only when absolutely necessary. Never move an injured person unless there is a fire or when explosives are involved. The major concern with moving an injured person is making the injury worse, which is especially true with spinal cord injuries. The American Red Cross suggests that if you must move an injured person, try to drag him or her by the clothing around the neck or shoulder area. If possible, drag the person onto a blanket or large cloth and then drag the blanket. For more information, see www.redcross.org.

Tricks of the Trade

✓ Here are some more tips for accident prevention when using scaffolding.

1. Inspect the scaffold after it is erected and each day during its use on the site.
2. Rope off the area underneath the scaffold.
3. To mount the scaffold, do not climb the braces; use a ladder.
4. Keep tools and materials away from the edge of scaffolds, so that they cannot be knocked off onto people working below.
5. Avoid an off-balance position when pulling, pushing or prying—especially when working at heights.
6. Rolling scaffold units should be no higher than four times their narrowest base measurement.
7. Always keep the casters of rolling scaffolding locked when it is not being moved. Never ride on rolling scaffolds.

Plans, Specifications, and Documentation

To understand programmable logic controllers, practice reading ladder diagrams. A ladder diagram may be an important tool for troubleshooting. They also may provide a map that

links hardware logic to programmed logic. The following example is a ladder diagram of a basic start/stop push-button control circuit for starting and stopping a motor.

In this diagram, the rungs of the ladder circuit are connected between the leads L_1 and L_2 from the control power supply. The stop button is normally closed and the start button is normally open. Depressing the start button results in power delivery to the motor starter coil, which pulls in the relays in the motor feeder. The relay M_1 seals in the circuit so the motor continues to operate after the start switch is released.

The above ladder diagram can be "mapped" onto a comparable programmable logic controller. Each component of the ladder diagram will become either hardware or software components that fall under the category of "inputs," "program" or "outputs." For example, the start/stop push button station provides one of two input options for the PLC. The PLC will then send an output signal to the motor starter coil. The program provides the "wiring" between the inputs and the outputs and each input and output will receive a unique address. Try drawing some of your own ladder diagrams and see if you can "map" them onto PLCs.

IEC NATIONAL PRIDE

Electrical Curriculum

Year Four Student Manual

Lesson 409 – Advanced Controls Lab #2

Purpose

To allow the student to use lab facilities to practice, experiment, and apply the knowledge learned in the previous two lessons.

Objectives

By the end of this lesson, you should be able to:

409-1   Install and connect photoelectric and proximity controls
409-2   Hook up and use a programmable controller

Content

Knowledge and Skills

This lesson is designed to instruct you in the following:

- Installing and connecting photoelectric and proximity controls.
- Using programmable controller applications multiplexing.

Professionalism

- Ability to plan before beginning work
- Ability to identify and use proper equipment and tools
- Willingness to organize work to make it as functional as possible
- Ability to communicate with installation team, including project managers

Advantages and Disadvantages of a PERT Chart

The advantages of a PERT chart include:

- It clearly shows relationship among activities
- It is easily understood by someone outside the project
- Deadlines and delegations can be specifically indicated with each step

The disadvantages of a PERT chart include:

- In order for the PERT chart to be useful, project tasks have to be clearly defined as well as their relationships to each other.
- The PERT chart does not deal very well with task overlap. The PERT chart assumes the following tasks begin after the prior task ends.
- The PERT chart is only as good as the time estimates that are entered by the project manager or designer.

Relevant Tools and Equipment

- Basic hand tools
- Meters

Code References

- None

Homework

1) Reading Assignment

- None

Advanced Controls Lab #2 (Student Manual) Lesson 409-3

2) Key Terms

Review the terms used in Lesson 407 and 408.

3) Practice Exercises

After experimenting and understanding photoelectric and proximity sensors, have students using these as various input devices for the programmable controller.

Study Tips

▶ When performing lab activities, you will sometimes run into problems. When you do, immediately ask for help. Don't get too far along before you realize that what you are doing is wrong or, possibly, dangerous.

▶ The instructor is there during lab classes to help you learn to perform your work in the proper manner. Help him do his job, by asking questions and checking out procedures as you go through the steps of the lab.

Tool Box Talk

Safety

✗ The American Red Cross suggests performing the Heimlich maneuver on choking victims. The following steps are suggested. Ask the victim to cough, speak, or breathe. If the victim can do none of these things, stand behind the victim and locate the bottom rib with your hand. Move your hand across the abdomen to the area above the navel then make a fist and place your thumb side on the stomach. Place your other hand over your fist and press into the victim's stomach with a quick upward thrust until the food is dislodged. For more information, see www.redcross.org.

Tricks of the Trade

✓ Control systems in industry may be open-loop systems, in which no feedback is provided, or closed-loop systems, in which feedback concerning changes in the controlled system is received. Most industrial systems are open-loop, such as a stepper motor in a milling machine. Once the machine operator selects the desired dimensions, and a computer converts the dimensions to a digital number representation, the motor turns causing the table to move. In a closed-loop system, adjustments to the process can be made based on parameter readings by sensors before a final result. This means sensed parameters (e.g. speed or temperature) still must be converted to a useful and observable signal for the operator to read. Note also that either positive or negative feedback indicates an error since only the value zero indicates that the actual value, when subtracted from the desired value, is the same as the desired value.

Plans, Specifications, and Documentation

The *National Electrical Code*® was originally developed to prevent fires caused by electricity. Because of resistance in a wire, current flow results in a small amount of voltage drop and this produces heat. The heat generated in wires and equipment can break down insulation over time and allow sparks to escape. This sequence causes most electrical fires. The heat generated by current must be controlled to prevent the fires. Excessive heat can also cause problems for automated controls. Be aware of temperature in electrical installations, both as heat generated by electrical circuits and as a factor in the environment (referred to as ambient temperature).

IEC PRIDE NATIONAL

Electrical Curriculum

Year Four
Student Manual

Lesson 410 – Mid-Term Review and Exam

Purpose

To review Lessons 404 through 409. Prepare students to take the Mid-Term Exam.

Objectives

By the end of this lesson, you should be able to:

404-1   Utilize the basic printed circuit (PC) board and its main components
404-2   Analyze semiconductor theory and its relation to semiconductor devices
404-3   Categorize N-type and P-type material
404-4   Classify rectification systems
404-5   Differentiate among various types of diodes
404-6   Explain the theory, operation, and use of various other solid-state devices

405-1   Organize the operations and functions of various electromechanical relays
405-2   Utilize proper contact arrangement terminology
405-3   Critique the operations and functions of solid-state relays
405-4   Select and install the proper relay for an application
405-5   Classify the advantages and disadvantages of different types of relays

406-1   Experiment, hook up and practice using solid-state control devices
406-2   Experiment, hook up and understand the operation of electromechanical and solid state relays

407-1   Evaluate when and how to use a photoelectric control in a control circuit
407-2   Explain how the different types of photoelectric controls function
407-3   Discriminate among the different types of proximity switches and how they function
407-4   Explain the Hall effect

408-1   Explain the uses of the programmable controller
408-2   Relate the functions of the parts of a programmable controller
408-3   Identify I/Os and write a simple program for a programmable controller

408-4 Categorize programmable controller applications
408-5 Identify the advantages of using multiplexing for specific applications

409-1 Install and connect photoelectric and proximity controls
409-2 Hook up and use a programmable controller

Content

Knowledge and Skills

This lesson is designed to instruct you in the following:

- Different types of solid-state devices.
- Electromechanical and solid state relays.
- Photoelectric and proximity controls.
- Programmable controllers.

Professionalism

- Ability to plan before beginning work
- Willingness to organize work to make it as functional as possible
- Ability to communicate with installation team, including project managers

The Pareto Principle

Another fundamental organizational is the Pareto Principle, more commonly known as the "80-20 Rule." The "80-20 Rule" was coined by Victor Pareto, an Italian economist and sociologist who studied the ownership of land in Italy at the turn of the 20th Century. In this study, he discovered that more than 80 percent of all the land in Italy at that time was owned by less than 20 percent of the people. And as he continued his studies, he found that this principle held true in everything he measured, including money. If this principle is true, then less than 20 percent of the work force produces more than 80 percent of the work that is done, no matter what kind of work is being done—sales, manufacturing, management, etc.

This rule can also be applied to efficiency and productivity in the workplace. Take a look at your "to do" list of projects and activities. According to the "80-20 Rule," about 20 percent of these projects and activities are producing 80 percent of the results coming from your work. And, most of the time, the first 20 percent of the time you spend working on a project will produce 80 percent of the rewards you get from that project.

Mid-Term Review and Exam (Student Manual) Lesson 410-3

Relevant Tools and Equipment

- Hand tools
- Soldering iron and solder
- Alligator clip leads
- Digital or analog Volt-Ohm meter
- Oscilloscope
- Continuity meters and beepers

Code References

- None

Homework

1) Reading Assignment

- Review all your homework assignments for Lessons 404-409.

2) Key Terms

Review all the key terms for Lesson 404-409.

3) Practice Exercises

- None

Mid-Term Review and Exam (Student Manual) Lesson 410-4

Study Tips

▶ These are a few hints to remember when you take tests like the Mid-Term Exam:

A. Be sure that you understand the instructions.
B. Read all the questions first. Then answer the easiest questions first. This will help build your confidence and ensure that you get credit for the easy questions.
C. Don't struggle with a question and waste your time. Go on and come back to it later.
D. Make sure that you allow enough time to finish all the questions.
E. If you are unsure of an answer, put down what you think is the best answer and come back to it later. Something in the rest of the test may trigger the correct answer for you.

Reduced Voltage Starters (Student Manual) Lesson 411-1

| *I E C* **PRIDE** *NATIONAL* | **Electrical Curriculum**

Year Four
Student Manual |

Lesson 411 – Reduced Voltage Starters

Purpose

To familiarize students with the theory of reduced voltage starters and the various techniques used in reduced voltage motor starters.

Objectives

By the end of this lesson, you should be able to:

411-1   Determine the reasons for reduced voltage starting in AC and DC motors
411-2   Explain how a primary resistor starter works
411-3   Explain how an autotransformer starter works
411-4   Explain how a part-winding starter works
411-5   Explain how a wye-delta starter works
411-6   Explain how a solid-state starter works

Content

Knowledge and Skills

This lesson is designed to instruct you in the following:

- Why do we use reduced voltage starters?
  - Voltage/current relationships
  - Power source
  - Load requirements
  - Electrical environment
  - DC motor construction
  - AC motor construction

- What is a primary resistor starter and how does it work?
  - Function and operation
  - Control circuitry

Copyright © 2008 by the Independent Electrical Contractors, Inc. All rights reserved.

- What is an autotransformer starter and how does it work?
  - Function and operation
  - Control circuitry
- What is a part-winding starter and how does it work?
  - Function and operation
  - Control circuitry
- What is a wye-delta starter and how does it work?
  - Theory
  - Function and operation
  - Control circuitry
- What is a solid-state starter and how does it work?
  - Theory
  - Function and operation
  - Control circuitry
  - Reduced voltage starter comparisons
- Starting method comparison
- Troubleshooting

Professionalism

- Ability to manage tasks, resources and costs effectively
- Ability to plan and organize work projects

Project Management

Project management is a system for managing tasks, resources and costs effectively. Managing projects means keeping scope, schedule and resources in balance. The scope of the project is the number of tasks required to get the project completed. The schedule refers to how you keep track of the time and sequence of each task in order to complete the project on time. Resources are the people, equipment, and materials need to do the project. Since people, equipment, and materials cost money, you need to keep track of these to make sure that the project is within budget.

Good project management helps you succeed. Everyone on the team will benefit from improved coordination and use of resources. Project management helps everyone have a sense that everything is under control.

Reduced Voltage Starters (Student Manual) Lesson 411-3

Relevant Tools and Equipment

- Basic hand tools

Code References

- Article 430

Homework

1) Reading Assignment

- Read *Electrical Motor Controls for Integrated Systems*, Chapter 17, pp. 485-508.

2) Key Terms and Formulas

Using either the *Illustrated Dictionary for Electrical Workers*, the Glossary in the Appendix or your textbooks, write definitions for each of these terms before doing your homework:

- Autotransformer starting
- Increment Current
- Part-winding starting
- Primary resistor starting
- Reduced voltage
- Solid state starting
- Wye-delta starting

Formulas

- $I = \dfrac{E}{R}$

- $S_{rpm} = \dfrac{120 \times F}{P}$

- $I = \dfrac{E - C_{EMF}}{R}$

Copyright © 2008 by the Independent Electrical Contractors, Inc. All rights reserved.

Reduced Voltage Starters (Student Manual) Lesson 411-4

3) Practice Exercises

Answer the following questions:

- Complete Tech-Chek 17 and do the worksheets for Chapter 17 in the *Workbook for Electrical Motors Controls for Integrated Systems*.

Study Tips

▶ In this lesson, you are asked to distinguish among five types of reduced voltage starters. It may help you to develop a chart listing the characteristics of each and how each works. This will help you to see the differences and will help you to remember these differences for future reference.

Tool Box Talk

Safety

✗ There are a many different types of burns. They can be thermal burns, chemical burns, electrical burns or contact burns. Each of the burns can occur in a different way, but treatment for them is very similar. For thermal, chemical or contact burns, the American Red Cross suggests the following steps for first aid. The first step is to run cold water over the burn for a minimum of 30 minutes. If the burn is small enough, keep it completely under water. Flushing the burn takes priority over calling for help. Flush the burn FIRST. If the victim's clothing is stuck to the burn, don't try to remove it. Remove clothing that is not stuck to the burn by cutting or tearing it. Cover the burn with a clean, cotton material. If you do not have clean, cotton material, do not cover the burn with anything. Do not scrub the burn and do not apply any soap, ointment, or home remedies. Also, don't offer the burn victim anything to drink or eat, but keep the victim covered with a blanket to maintain a normal body temperature until medical help arrives. For more information, see www.redcross.org.

Reduced Voltage Starters (Student Manual) Lesson 411-5

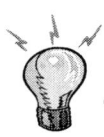

Tricks of the Trade

✓ A motor sitting at rest has a different electrical characteristic than one already rotating. To get the motor to overcome standstill requires supplying the current not only for the rotational torque but also for the magnetic fields. The impedance of a motor, as seen by the electrical source, is much lower once the motor is turning. Lower impedance means that more current will flow when connected to the same voltage then during the steady-state condition. In a motor, this "inrush" current is often six to ten times higher than current present while running. Motors driven by electronic controls, such as adjustable speed drives, have helped with problems caused when voltage drops quickly and steadily increases. Nevertheless, in older industrial facilities motor starts are probably the leading cause of "sags" that originate on the load side, or from within the facility.

Plans, Specifications, and Documentation

A motor nameplate is an important source of information about the circuit specifications for a particular motor. The specifications on the nameplate are based on *NEC®* Article 430. There are several important pieces of information required for wiring the circuit.

1. Horsepower

2. Phase

3. Voltage

4. Full-load current

5. Temperature rise above ambient or service factor (to prevent overheating).

The nameplate will also give the motor code letter and design letter. An installer also needs to consider the physical environment, the location of the installation, type of controller desired, and the type of load that will be powered (e.g. easy-starting or hard starting load).

Accelerating and Decelerating Methods (Student Manual) Lesson 412-1

| *I E C* **PRIDE** *NATIONAL* | Electrical Curriculum

Year Four
Student Manual |

Lesson 412 – Accelerating and Decelerating Methods

Purpose

To teach the student how motors are be braked and various techniques used in motor speed control.

Objectives

By the end of this lesson, you should be able to:

412-1   Distinguish among various motor braking techniques
412-2   Explain multispeed motors and their connections
412-3   Utilize various methods of controlling motor speed

Content

Knowledge and Skills

This lesson is designed to instruct you in the following:

- How is motor braking accomplished and what factors must be taken into consideration in braking a motor.
  - Braking
  - Friction brakes
  - Braking torque
  - Plugging
  - Electric braking
  - Dynamic braking

Copyright © 2008 by the Independent Electrical Contractors, Inc. All rights reserved.

- Characteristics to consider with respect to speed control.
  - Load requirements
  - Force and work
  - Torque
  - Horsepower
  - Constant torque-variable horsepower
  - Constant horsepower-variable torque
  - Variable torque-variable horsepower
- Controlling the speed of AC and DC motors
  - DC vs. AC motors
  - Compelling circuit logic
  - Accelerating circuit logic
  - Decelerating circuit logic
  - Frequency change
  - Variable frequency drives

Professionalism

- Ability to plan and organize work materials
- Ability to manage tasks, resources and costs effectively

Accelerating and Decelerating Methods (Student Manual) Lesson 412-3

Planning in Project Management

Planning is an important part of project management. Planning allows you to establish what is required to successfully complete the project. Planning helps you make sure that quality standards are met, timelines are adhered to and costs are kept within budget. What follows are some steps for planning:

- Establish the project goal,
- Choose the basic strategy for achieving that goal,
- Break down the project into steps or tasks,
- Determine the performance standards for each step or task,
- Determine how much time is required to complete each task,
- Determine the proper sequence for completing each task,
- Determine the cost of each task,
- Decide who on your team will need to do each task,
- Develop a work plan that outlines all the information from the steps above.

What are tasks? Why break your project into tasks? A project task is a related set of work steps that help to get a project done. A list of tasks can be short and simple or long and complex, depending on the size and goal of the project. There are several reasons for organizing your project into smaller work units called tasks:

- By breaking work into tasks, you can put the in work in your project in the most logical sequence for completion. This can help you determine the best schedule for your project.
- Tracking the progress of tasks is a way to assess and control the work done on a project.
- By breaking the work into tasks, you can determine the skills you need to complete the work on a project and you can determine how many people will be required to do the work.
- Tasks allow you to communicate the work that needs to be done to other team members.
- A task list can be used to negotiate changes in work sequences during the project.
- Breaking the work into tasks ensures that all work sequences are identified and understood.

Relevant Tools and Equipment

- None

Code References

- None

Homework

1) Reading Assignment

- Read *Electrical Motor Controls for Integrated Systems,* Chapter 18.

2) Key Terms and Formulas

Using either the *Illustrated Dictionary for Electrical Workers,* the Glossary in the Appendix or your textbooks, write definitions for each of these terms before doing your homework:

- Accelerating circuit logic
- Braking
- Compelling circuit logic
- Decelerating circuit logic
- Dynamic Braking
- Multispeed
- Plugging
- Torque

Formulas

- $T = \dfrac{5252 \times HP}{rpm}$

- $HP = \dfrac{I \times E \times eff}{746}$

- $HP = \dfrac{rpm \times T}{5252}$

- $V/HZ = \dfrac{V}{HZ}$

- $PD_m = \dfrac{PD_d \times N_d}{N_m}$

- $rpm_{syn} = \dfrac{120 \times f}{N_p}$

- $Work = Distance \times Force$

Accelerating and Decelerating Methods (Student Manual) Lesson 412-5

3) Practice Exercises

Answer the following questions:

- Complete Tech-Chek 18 and the worksheets for Chapter 18 in the *Workbook for Electrical Motors Controls for Integrated Systems*.

Study Tips

▶ This is the first lesson this year that makes use of formulas. While they are relatively simple formulas, make sure that you understand them and can use them to determine torque, power and horsepower for multispeeed motors. If you are not sure that you can use them appropriately, ask your instructor for extra problems for practice.

Tool Box Talk

Safety

✗ If someone has received an electrical burn, the American Red Cross recommends the following first aid steps. Note that these steps are different than those for other kinds of burns. First of all, don't touch a victim who has been in contact with electricity unless you are clear of the power source. If the victim is still in contact with the power source, electricity will travel through the victim's body and electrify you when you reach to touch. Once the victim is clear of the power source, your priority is to check for any airway obstruction, and to check breathing and circulation. Administer CPR if necessary. Once the victim is stable, begin to run cold water over the burns for a minimum of 30 minutes. Don't move the victim and don't scrub the burns or apply any soap, ointment, or home remedies. After flushing the burn, apply a clean, cotton cloth to the burn. If cotton is not available, don't use anything. Keep the victim warm and still and try to maintain a normal body temperature until medical help arrives. For more information, see www.redcross.org.

Tricks of the Trade

✓ Motor speed control is accomplished using feedback of supply voltage, speed, or motor current. In the case of automated controls, a microprocessor adjusts supplied motor voltage, frequency, or current to obtain a zero error from the requested speed. Motor acceleration is controlled to assure either a rate of acceleration based on time or a current limit usually specified at 125% of maximum steady-state running current.

Copyright © 2008 by the Independent Electrical Contractors, Inc. All rights reserved.

Plans, Specifications, and Documentation

The National Electrical Manufacturers Association (NEMA) has established important specifications for motor design. The design letter on the nameplate of a motor will give the NEMA specification. The most common type of motor in use is the design B motor. NEMA has also established a size numbering system for horsepower ratings of electric motor starters. Voltage ratings and motor horsepower ratings are given for single-phase and three-phase motor starters. Get to know these designations. They provide installers with a lot of information for the correct design of motor branch circuits.

Advanced Controls Lab #3 (Student Manual) Lesson 413-1

| *I E C* **PRIDE** *NATIONAL* | **Electrical Curriculum**

Year Four
Student Manual |

Lesson 413 – Advanced Controls Lab #3

Purpose

To allow the student to experiment, hook up, and practice using various reduced voltage starters.

Objectives

By the end of this lesson, you should be able to:

413-1   Practice using various reduced voltage starters
413-2   Employ the brakes and control the speed of various motors

Content

Knowledge and Skills

This lesson is designed to instruct you in the following:

- Installing and using reduced voltage starters.
- Using brakes and controlling the speed of various motors.

Professionalism

- Work as a life-long learner to keep abreast of constant changes in the industry
- Ability to plan and organize work materials
- Ability to manage tasks, resources and costs effectively

How Long Does It Take to Get Things Done?

When planning a project, there are several characteristics of good task management that it may help to keep in mind. First, a task should be clear and state exactly what should be done. Secondly, all work within a task should occur in sequence without gaps for other tasks in between. For instance, "Framing and wiring the library" is two tasks, not one. Thirdly, most steps in a task should use the same team members. If it requires a different set of skills and a different team member, it is probably a different task. Fourthly, a task that involves multiple people usually goes smoothly if all the people working on the task are at the job site and within walking distance of each other. Lastly, in scoping out the size of a task, you can use the 8/80 and the reporting period rule. The 8/80 rule means that you keep your tasks between eight hours (one work day) and 80 hours (10 work days). The reporting period rule states that no task should be longer than the standard reporting period. Thus, if you make weekly reports, no task should be longer than a week. This helps you keep tasks on the same schedule as your reporting cycle.

Relevant Tools and Equipment

- Basic hand tools
- Multimeters

Code References

- Article 430

Homework

1) Reading Assignment

- None

2) Key Terms and Formulas

Review the terms used in Lessons 411 and 412.

Advanced Controls Lab #3 (Student Manual) Lesson 413-3

3) Practice Exercises

Complete the following activity:

- Students should work in small groups and rotate to the various types of starters and motors, so that everyone has a chance to examine each type of starter and motor.

Study Tips

▶ Ask yourself the following questions during each lab: "So what? Why do I need to learn this?" "What did I learn from this lab?" "Have I done this on the job?" "How did what I did on the job differ?" "How was it the same?" "How much do I think I will use this skill on the job?" If you have trouble answering any of these questions you need to discuss them with your instructor.

Tool Box Talk

Safety

✘ Heat exhaustion and heat stroke are two different things, although they are commonly confused as the same condition. Heat exhaustion can occur anywhere there is poor air circulation, such as around an open furnace or heavy machinery, or even if the person is poorly adjusted to very warm temperatures. The body reacts by increasing the heart rate and strengthening blood circulation. Simple heat exhaustion can occur due to loss of body fluids and salts. The symptoms are usually excessive fatigue, dizziness and disorientation, normal skin temperature but a damp and clammy feeling. To treat heat exhaustion, the American Red Cross recommends moving the victim to a cool spot and encouraging drinking of cool water and rest. For more information, see www.redcross.org.

✘ Heat stroke is much more serious and occurs when the body's sweat glands have shut down. Some symptoms of heat stroke are mental confusion, collapse, unconsciousness, fever with dry, mottled skin. According to the American Red Cross, a heat stroke victim will die quickly, so don't wait for medical help to arrive—assist immediately. The first thing you can do is move the victim to a cool place out of the sun and begin pouring cool water over the victim. Fan the victim to provide good air circulation until medical help arrives. For more information, see www.redcross.org.

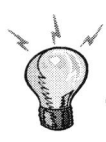

Tricks of the Trade

✓ Motor controllers are usually called starters or drives. Reduced-voltage starters assure that less than rated voltage is applied to an AC motor during starting to protect motor windings and to prevent high current draw. Current systems usually use solid-state controls although staging resistors and autotransformers are also used.

Plans, Specifications, and Documentation

The *NEC*® provides information that will assist you in determining how to design a motor branch circuit. Get to know these Code areas and use them.

1. Markings on motors and what they mean—*NEC*® 430.7
2. Motor circuit conductors—*NEC*® 430.21 through 430.29
3. Specify overload protection—*NEC*® 430.31 through 430.44
4. Short circuit protection—*NEC*® 430.52 through 430.58 and *NEC*® 240.6
5. Specify disconnecting means—*NEC*® 430.101 through 430.113

Preventative Maintenance and Troubleshooting (Student Manual) Lesson 414-1

I E C
PRIDE
NATIONAL

Electrical Curriculum

**Year Four
Student Manual**

Lesson 414 – Preventative Maintenance and Troubleshooting

Purpose

To teach students how to troubleshoot circuits and employ preventative maintenance techniques.

Objectives

By the end of this lesson, you should be able to:

414-1   Specify preventative maintenance techniques and programs
414-2   Demonstrate proper troubleshooting techniques
414-3   Evaluate why motors fail

Content

Knowledge and Skills

This lesson is designed to instruct you in the following:

- What is preventative maintenance?
  - Inspection
  - Cleaning
  - Tightening
  - Adjustment
  - Lubrication
- How do you troubleshoot?
  - Test instruments
  - Troubleshooting procedures
  - Troubleshooting techniques
  - Troubleshooting power circuits
  - Troubleshooting motors

- ➢ Remarking motor connections
- ➢ Troubleshooting circuit components
- Understanding why motors fail.
  - ➢ Loading
  - ➢ Voltage
  - ➢ Unbalance
  - ➢ Single-phasing
  - ➢ Duty cycle
  - ➢ Temperature
  - ➢ Ventilation

Professionalism

- Work as a life-long learner to keep abreast of constant changes in the industry
- Ability to plan and organize work materials
- Ability to manage tasks, resources and costs effectively
- Ability to set goals and see them through to completion

Preventative Maintenance and Troubleshooting (Student Manual) Lesson 414-3

Project Goals

Every project has three primary goals:

1. To create something,
2. To complete it within a specific budgetary framework,
3. To finish within an agreed upon schedule.

Beyond these goals are the other goals that must be specified that actually define the project. To distinguish primary goals from other project goals, the other goals are sometimes referred to as objectives. But goals and objectives are usually different words for the same thing. No matter what they are called, they should meet the following criteria:

- Goals must be specific. Your goals should be so clear that another project manager should be able to take over for you, if need be.
- Goals must be realistic. Your goals must be within the realm of possibility.
- Goals must have a time in which they must be accomplished. Projects must have a definite finish date or they may never be completed.
- Goals must be measurable. You must be able to measure your success at meeting your goals. These results are called deliverables—the results of the project.
- Goals must be agreed upon. You, your boss and your customer must agree on the goals before you take any further steps toward planning the project.

Responsibility for achieving the goals must be identified. Although you as project manager bear the brunt of responsibility for the overall success of the project, others may be responsible for pieces of goals. The people responsible for the goals must be identified and be willing to accept responsibility before the project proceeds. It is best to have the sign-off of all the major players before you begin the project.

Relevant Tools and Equipment

- Basic hand tools
- Multimeters
- Amprobe

Code References

- None

Copyright © 2008 by the Independent Electrical Contractors, Inc. All rights reserved.

Homework

1) Reading Assignment

- Read *Electrical Motor Controls for Integrated Systems*, Chapter 18.

2) Key Terms and Formulas

Using either the *Illustrated Dictionary for Electrical Workers*, the Glossary in the Appendix or your textbooks, write definitions for each of these terms before doing your homework:

- Phase loss
- Phase reversal
- Phase unbalance

Formulas

- $V_u = \dfrac{V_d}{V_a} \times 100$

3) Practice Exercises

Answer the following questions:

- Complete Tech-Chek 18 and the worksheets for Chapter 18 in the *Workbook for Electrical Motors Controls for Integrated Systems*.

Study Tips

▶ One of the key aspects of an electrician's work is to troubleshoot and repair motor control circuits. Spend some time learning the common troubles, causes, and remedies for motor controls. It will be well worth your time.

▶ Remember: When troubleshooting or performing preventative maintenance always think safety. Remember to use lockout-tagout procedures as required by OSHA.

Tool Box Talk

Safety

✗ In cases of poisoning, the American Red Cross recommends that you take the following first aid steps. The first thing to do is get the victim away from the poison. Then provide treatment appropriate to the form of the poisoning. If the poison is in solid form, such as pills, remove it from the victim's mouth using a clean cloth wrapped around your finger. Don't try this with infants because it could force the poison further down their throat. If the poison is a gas, you may need a respirator to protect yourself. After checking the area first for your safety, remove the victim from the area and take the victim to fresh air. If the poison is corrosive to the skin, remove the clothing from the affected area and flush with water for 30 minutes. Take the poison container or label with you when you call for medical help because you will need to be able to answer questions about the poison. Try to stay calm and follow the instructions you are given. If the poison is in contact with the eyes, flush the victim's eyes for a minimum of 15 minutes with clean water. For more information, see www.redcross.org.

Tricks of the Trade

✓ If you are having trouble understanding mechanical or electrical blueprints take a copy of the plans into the building and compare what has been installed with what shows on the plans. This will help you understand the installations of other trades. It will also give you a better understanding of the factors that can cause conflicts between plumbing, mechanical, and electrical installations.

Plans, Specifications, and Documentation

Always consult the manufacturer's recommended maintenance schedule for components and equipment. Whether you make repairs or just do routine maintenance, don't forget to document what you have done. If any changes were made to electrical or mechanical equipment, they must be noted on the appropriate shop drawings. This is particularly so for changes made to the electrical wiring of machinery.

Advanced Controls Lab #4 (Student Manual) Lesson 415-1

I E C NATIONAL

Electrical Curriculum

**Year Four
Student Manual**

Lesson 415 – Advanced Controls Lab #4

Purpose

To allow the student to use lab facilities to practice, experiment, and apply the knowledge learned in the previous two lessons.

Objectives

By the end of this lesson, you should be able to:

415-1   Troubleshoot motor control system

Content

Knowledge and Skills

This lesson is designed to instruct you in the following:

- Practice troubleshooting of motor control systems.

Professionalism

- Work as a life-long learner to keep abreast of constant changes in the industry
- Ability to plan and organize work materials
- Ability to manage tasks, resources and costs effectively
- Ability to set goals and see them through to completion

Copyright © 2008 by the Independent Electrical Contractors, Inc. All rights reserved.

Establishing Project Goals

It is easy to establish goals for your project. In fact, usually it is so easy that you'll find yourself with more goals than you can handle. Establishing goals takes time to ensure that you aren't biting off more than you can chew. Here's what to do to establish good overall project goals:

- Make a list of the project goals. At this point, don't rule anything out. Just make a list and check it twice.

- Study the list and eliminate anything that has no direct bearing on the project.

- Eliminate anything that is really a step in meeting the goals and is not a goal for the end result of the project.

- Study the list again. Make sure each goal meets all the relevant criteria listed above. Now determine whether all these goals are doable within one project. Look for goals that aren't directly related to the project. Cross them out. They may be really important, but may apply to another project. Keep them for that project. If the items you removed form the list are important, you should consider alerting your management so they can decide if any action needs to betaken. You should be prepared to justify why they don't fit in your project.

Your task as project manager is to run the project as smoothly as possible with no additional time, bodies or money. If you take time to develop a good, realistic set of goals, this is the result you will get.

Relevant Tools and Equipment

- Basic hand tools
- Meters

Code References

- Article 430

Advanced Controls Lab #4 (Student Manual) Lesson 415-3

Homework

1) Reading Assignment

- None

2) Key Terms and Formulas

Review the terms and formulas used in Lesson 414.

3) Practice Exercises

Completed lab activities as assigned.

Study Tips

▶ Trouble shooting is a form of problem solving. When troubleshooting it may help to remember some tips about problem solving:

- The first and most important step in solving a problem is to understand the problem.
- Next you need to devise a plan to solve the problem. Identify which skills and techniques you have learned that can be applied to troubleshoot the problem.
- Carry out the plan
- Look back—Did your troubleshooting solve the problem? If it did, review your method of solution so that you will be able to recognize and solve similar problems. If you didn't solve the problem, you should start over again with the first step.

Tool Box Talk

Safety

✘ The American Red Cross suggests the following FIRST AID KIT CHECKLIST. In order to administer effective first aid, it is important to maintain adequate supplies in each first aid kit. First aid kits can be purchased commercially already stocked with the necessary supplies, or a first aid kit can be made by including the following items. For more information, see www.redcross.org.

- Adhesive bandages: available in a large range of sizes for minor cuts, abrasions and puncture wounds

- Butterfly closures: these hold wound edges firmly together.

- Rolled gauze: these allow freedom of movement and are recommended for securing the dressing and/or pads. These are especially good for hard-to-bandage wounds.

- Nonstick Sterile Pads: these are soft, superabsorbent pads that provide a good environment for wound healing. These are recommended for bleeding and draining wounds, burns, and infections.

- First Aid Tapes: Various types of tapes should be included in each kit. These include adhesive, which is waterproof and extra strong for times when rigid strapping is needed; clear, which stretches with the body's movement, good for visible wounds; cloth, recommended for most first aid taping needs, including taping heavy dressings (less irritating than adhesive); and paper, which is recommended for sensitive skin and is used for light and frequently changed dressings.

- Items that also can be included in each kit are tweezers, first aid cream, thermometer, an analgesic or equivalent, and an ice pack.

Advanced Controls Lab #4 (Student Manual) Lesson 415-5

Tricks of the Trade

✓ An electrician may have to do some heavy lifting from time to time. It is a good idea to refresh your knowledge of proper material handling and lifting techniques to use on the job.

1. When lifting, evaluate the size of the load and get help if it is too large for one person
2. Bring the object close to you, centering the weight over your feet.
3. Lift smoothly and avoid quick, jerky motions.
4. When carrying a heavy load, shift your feet instead of twisting your body.
5. To lift a load above waist height, rest it on a table or bench, shift your grip, and then lift again.
6. When carrying a heavy load, use two people when necessary. Plan in advance the route along which the load will be moved.
7. Do not let the object you are moving obstruct your vision. Always have a clear view of where you are going.
8. Carry conduit and other long objects on your shoulder.
9. Push or pull at waist height; avoid bending and twisting when pushing or pulling. Whenever possible, push instead of pulling.
10. Use appropriate material handling equipment whenever possible.

Plans, Specifications, and Documentation

In many communities, electricity and/or natural gas is delivered through buried systems. Digging safely and correctly is essential when working in areas with underground lines. Before you begin any project—digging fence posts, planting trees, excavating—that requires you to move earth, call your state's one-call center to mark your underground utility lines. In most states it is the law! It's a free service in most states and could save you considerable expense, avoid inconvenience, and possibly save your life.

Leadership (Student Manual) Lesson 416-1

I E C PRIDE NATIONAL

Electrical Curriculum

Year Four
Student Manual

Lesson 416 – Leadership

Purpose

To help students develop their leadership skills to effectively manage projects and build productive teams.

Objectives

By the end of this lesson, you should be able to:

416-1   Identify your personal leadership style
416-2   Develop project management skills
416-3   Develop skills in communicating effectively with others in a team setting
416-4   Develop skills in building successful work teams

Content

Knowledge and Skills

This lesson is designed to instruct you in the following:

- Leadership Style Inventory
- Project Management
- Team Building

Professionalism

- Ability to build effective work teams.
- Ability to plan and organize work materials
- Ability to manage tasks, resources and costs effectively
- Ability to set goals and see them through to completion
- Recognize and promote project quality

Copyright © 2008 by the Independent Electrical Contractors, Inc. All rights reserved.

Team Motivation

To increase motivation in your team members you should be willing to listen to their concerns with an open mind. The following activities can be helpful when you are trying to increase your team's motivation.

- Help and encourage team members to increase their job skills;
- Provide opportunities for them to express their concerns and ideas;
- Give regular feedback to them regarding their job performance. Notice and praise positive change;
- Give team members your time and attention. Develop a positive relationship with them;
- Encourage members to make decisions about how to complete their jobs whenever possible;
- Acknowledge the difficulties and challenges in their job—even those that can't change;
- Recognize and praise steps towards accomplishing team goals;
- Help create a fun work environment.

Relevant Tools and Equipment

- None

Code References

- None

Homework

1) Reading Assignment

- Read the Project Management and Team Building Handbook prior to class. Be prepared to participate in class activities.

Leadership (Student Manual) Lesson 416-3

2) Key Terms and Formulas

Using the Leadership Style Inventory, write the definitions for each of these terms before doing your homework.

- Leadership
- Productivity
- Motivation
- Rewards
- Communication
- Consensus
- Project Management

3) Practice Exercises

- In-class activity
- Leadership Style Inventory

Study Tips

▶ Remember, in taking an inventory like this one, there are no right or wrong answers. These are just your perceptions of your behavior. The aim of the inventory is to describe how you lead, not to evaluate your leadership ability.

▶ Last week's lesson was an Advanced Controls Lab. You should continue to review and study for your Semester Exam. Be sure to ask your instructor for help if there is some area that needs more review.

▶ You should be reviewing and studying for your Semester Exam. Be sure to ask your instructor for help if there is some area that needs more review.

Tool Box Talk

Safety

✘ When it comes to safety on the job, there is only one right thing to do and that is to follow correct safety procedures regardless of what others think or say. Resist the temptation to speed up your work by cutting corners on safety, even if others are doing it. In the long run it doesn't pay off, it leads to more accidents, lost work time, and higher insurance rates. All of these affect your company's productivity.

✘ Doing the right thing also means doing something about unsafe conditions you see around you. All it takes is a minute or a word to your supervisor. When you report unsafe conditions or procedures that could lead to an accident, it shows that you care about the people you work with. In the long run, doing the right thing will gain you respect. And it will encourage others who really want to be safe to join with you in doing the right thing.

Tricks of the Trade

✓ Ensure that your tools are in good condition. Don't use a damaged tool to perform a task. Dull tools can cause damage to both cables and installers. Always use the correct tool for the task at hand.

✓ Lineman's scissors or electrician snips can also be used to remove a cable jacket. Always use extreme caution if using scissors to remove a cable jacket. This tool is very sharp and can easily cut into the cable conductors and nick the insulation on the wires. These scissors should not be the first choice as a cable-stripping tool.

✓ Pay careful attention when you are using a potentially dangerous tool. Devote your whole attention to what you are doing to make sure that you do not accidentally cut yourself or the cable you are working with. When working in an occupied area, take extra precaution to secure the work area.

Plans, Specifications, and Documentation

- Know the contract documents. It is especially important that you understand the Scope of Work and the Construction Blueprints.

- The Scope of Work is sometimes a separate document that details the work to be accomplished. Most often on larger jobs, the Scope of Work is included in the construction Specifications or Project Manual. It is very important to understand where your trade's responsibility begins and ends and another trade's responsibilities take over. With telecommunications installers, this is very important when coordinating with other contractors.

- Know the Construction Blueprints. Pay particular attention to notes on the drawings. These note many times will change or modify the general installation requirements. Don't pay attention to the electrical blueprints only. Know the mechanical plans, as well as the architectural and site plans.

Project Management and Team Building Handbook

Project Management and Team Building Handbook–2

Topic	Page Number

Introduction 3

Overview—Becoming a Supervisor 4
    - Developing your Leadership Style..5
    - Leadership Style Inventory 7

Building a Successful Team 9
    - A Successful Team 9
    - Team Productivity 11
    - How to Use Rewards..11

Team Communication 13
    - How to Use Team Communication 13
    - Barriers to Effective Communication/Conflict 14
    - Resolving Team Conflict..14
    - Building Consensus/Team Problem Solving 15

Project Management 16
    - Introduction to Project Management 16
    - Defining the Project and Tasks 16
    - Planning..17
    - Implementing the Project 18
    - Tracking Progress 22
    - Closing a Project 25

Introduction to This Handbook

This handbook was developed as a supplemental tool for students completing the IEC Electrical curriculum. Topics covered in this book include building a team and team supervision. Many people commonly recognize that the work involved in electrical services often requires a team approach to successfully complete it. As in many fields, a person highly skilled at his/her craft is often promoted to a supervisory position. However, these individuals may not have had the opportunity to develop the skills needed to be a successful supervisor of a team.

This handbook aims at providing information about those important topics of team building and supervision in an effort to help students develop these skills for their future. It is encouraged that students keep this handbook for future reference.

Overview—Becoming a Supervisor

Congratulations! You've just been promoted to a supervisory position. Now what? You are excited about the new opportunity, but also nervous about taking on the new challenges. You have many questions about how you'll do in your job. You might ask yourself:

- Am I really able to be a supervisor?
- What should I do if my workers start arguing?
- Will my workers accept me as their supervisor?

The answers to these questions can be boiled down to developing some key skills that will help you to become the type of supervisor you want to be. Suggestions for transitioning to the position of supervisor include:

- Use good communication skills. If you don't have them, start working on them right away. Some skills you want to have include giving feedback, active listening, conflict resolution, communicating clearly, and openness to differences. This handbook will cover communication skills in more detail in a later section.

- As a supervisor, you want to remain friendly with your workers, but being good friends may not always work. You want to appear fair to all the workers, not favoring one person over another. If you are good friends with one or more of your workers, it may be difficult to give them feedback about their job performance when things aren't going well.

- Don't act like "the big boss." You want to be approachable and sincere to your workers. Most people don't like it when their supervisors act arrogantly or superior.

- Challenge yourself to figure out how to motivate your workers. What can you do as a supervisor to keep them focused on their work? What will help them want to keep coming to work each day? If your workers are happy in their jobs, your job as a supervisor will be much easier.

- Learn to delegate work duties to others. Delegating work is not simply getting someone else to do a particular task. It is the opportunity for you to help a worker learn a new task, free up some of your time for other duties, and show your workers that you have confidence in them to complete a task.

- Realize that there will be differences among your workers. Each worker has different motivational level, skills, and life situations. As a supervisor, you want to be flexible while still appearing fair. What makes a technician "tick" and how can you as the supervisor tap into it? Often times observing your workers can give you some clues as to what motivates them to do a great job. Also, asking them directly can be very effective. Remember that what motivates one person may not motivate another. You may have to adjust how you do things by individuals, small groups, or by entire teams. The key question to ask yourself is "What will help me motive my team to do the best job possible?"

- Be willing to let go of things from your previous position. One sure way for a new supervisor to fail is trying to do everything. In your new position, you should start placing your trust in your workers to complete their jobs.

Project Management and Team Building Handbook–5

Developing Your Leadership Style

There are many leadership style inventory tools available on the market. As a way for you to assess your own skill level, we are providing the following tool for your use. This can help you evaluate what skills you may want to develop to improve as a supervisor.

Ways for Leaders to Behave

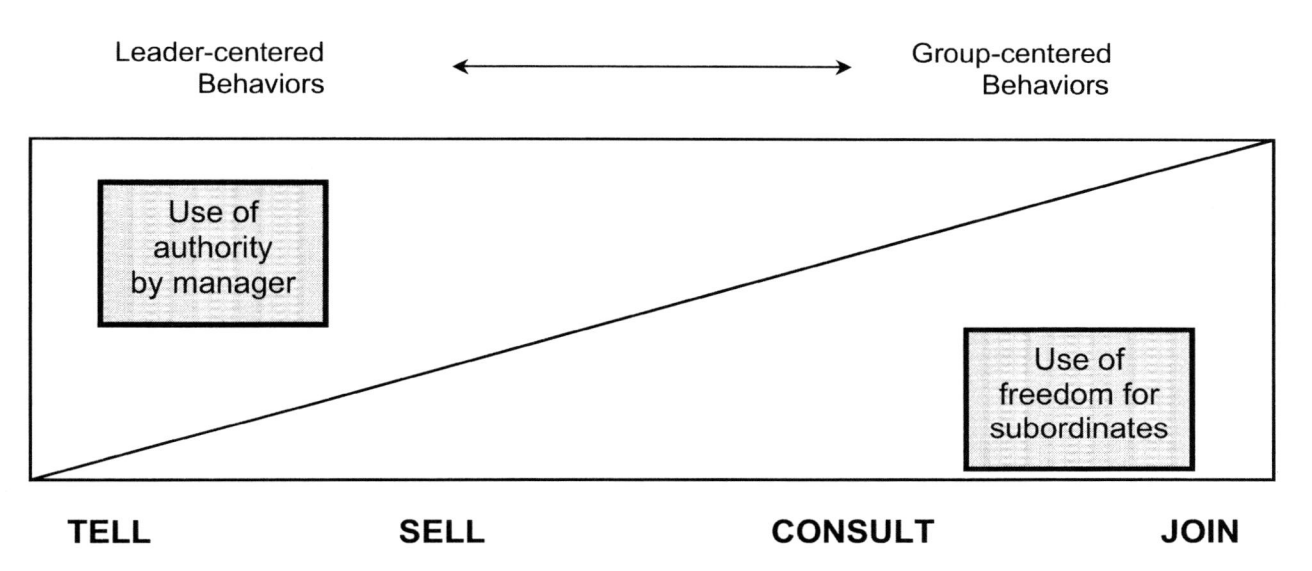

Leader-centered Behaviors ←——————→ Group-centered Behaviors

Use of authority by manager

Use of freedom for subordinates

TELL **SELL** **CONSULT** **JOIN**

*The Tannenbaum and Schmidt Leadership Model. J.F. Veiga, University of Connecticut.

LEADERSHIP STYLE

Whether you decide to be directive, supportive, challenging, or participative, or to delegate decision making to others, you are ultimately accountable for the results. The choice of your leadership style will depend on your own preference and on the willingness and readiness of your subordinates to assume more responsibility. It is most important for you to identify your usual or preferred style of leadership. Until you identify it, you cannot make the adjustments needed to succeed. To help you do this, complete the Leadership Style Inventory and then read the explanations of the four categories of leadership behavior.

- **TELL** A manager identifies a problem, considers alternative solutions, chooses one of them, and then reports this decision for implementation. He/she may or may not give consideration to what he/she believes his/her subordinates will think or feel about the decision. In any case, *he/she provides no opportunity for them to participate.*

- **SELL** As with TELL, a manager takes responsibility for identifying the problem and arriving at a decision. But rather than simply announcing it, one takes the additional step of *persuading one's subordinates to accept the decision.* One recognizes possible resistance and seeks to reduce resistance by one's actions.

- **CONSULT** A manager identifies a problem, *consults his/her subordinates for possible solutions*, and then makes the final decision. He/she recognizes the need to effectively cull from his/her subordinates their ideas to give them a sense of ownership and therefore commitment to the final decision as well as to discover other possible solutions her subordinates might know.

- **JOIN** A manager defines a problem and its limitations, and then *passes to the group* (including him/herself) *the right to make the final decision.* He/she feels her subordinates are capable of making decisions as good as or better than his/her own. He/she feels that human resources are best utilized by allowing them equal decision-making authority.

A democratic leader focuses on building a team or sharing each person's expertise among all the members. "From each according to knowledge, to each according to need" describes this leadership stance.

When should a leader be more directive? When tasks are ambiguous, when organizational policies are unclear, or when subordinates are unable or unwilling to take on responsibilities. Leadership should be supportive when subordinates work on stressful, frustrating, or dissatisfying tasks. A leader who provides challenge by being achievement-oriented will work best with subordinates performing nonroutine ambiguous tasks. It is possible that people who choose to do routine, repetitive jobs may have a type of personality that does not respond well to challenge. Participative leadership works best when subordinates prefer autonomy and self-control. We know that, when people can participate in the decision making that influences their jobs, they are more committed to the results of those decisions. When people can own a decision, they have a feeling of responsibility for its outcome.

Subordinates who have authoritarian personalities prefer a directive leader who is nonparticipative. Therefore, a manager must at all times be aware of, not only the types of tasks, but also of the types of employees with whom he/she is working. How do you know when an employee is an authoritarian type? When a person is rigid, has very little flexibility, does not like change, does not respond well to the unpredictable, and relates to rules rather than people's needs, he/she could be considered authoritarian.

LEADERSHIP ISSUES

"The world is divided into those who look and those who are looked at." If you are a leader in a power position, you are looked at. If you are a follower, or in a powerless position, you do the looking.

What does this mean? Leaders are visible. Their actions are constantly commented upon; their decisions are held up to public scrutiny; they have no hiding place. They are admired, hated, misunderstood, misinterpreted, sought after, avoided, tested, deferred to, and agreed with.

One of the gravest dangers threatening people in management positions is the lack of information. Subordinates, wanting to ingratiate themselves, may only report good news; hard truths can be kept from the top executive, who then acts on misinformation and lack of data. It is critical for an effective executive to keep seeking new data and to be open to unpleasant information, be it about his/her leadership style or the consequences of his/her decisions or policies.

Leaders are also subject to counterdependence from people who react negatively to anyone in authority. A woman may be more prone to elicit such counterdependence because of her gender. In this case, anything she says or does will be seen negatively—it has nothing to do with style or competence. It is important to differentiate between taking the blame and leaving it out there where it belongs, in the prejudicial attitudes of some subordinates.

Project Management and Team Building Handbook–7

Leadership Style Inventory*

This inventory was designed to assess your method of leading. As you fill out the inventory, give a high rank to those words that best characterize the way you lead and a low rank to the words that are least characteristic of your leadership style. You may find it hard to choose the words that best describe your leadership style because there are no right or wrong answers. Different characteristics described in the inventory are equally good. The aim of the inventory is to describe how you lead, not to evaluate your leadership ability.

Instructions

There are nine sets of four words listed below. Rank order each set of four words, assigning a 4 to the word which best characterizes your leadership style, a 3 to the word which next best characterizes your leadership style, a 2 to the next most characteristic word, and a 1 to the word which is least characteristic of you as a leader. *Be sure to assign a different rank number to each of the four words in each set.* Do not make ties. Then total the columns, using only the sets numbered below in the scoring section.

#				
1.	___ Forceful	___ Negotiating	___ Testing	___ Sharing
2.	___ Decisive	___ Teaching	___ Probing	___ Unifying
3.	___ Expert	___ Convincing	___ Inquiring	___ Cooperative
4.	___ Resolute	___ Inspirational	___ Questioning	___ Giving
5.	___ Authoritative	___ Compelling	___ Participative	___ Approving
6.	___ Commanding	___ Influential	___ Searching	___ Collaborating
7.	___ Direct	___ Persuasive	___ Verifying	___ Impartial
8.	___ Showing	___ Maneuvering	___ Analytical	___ Supportive
9.	___ Prescriptive	___ Strategical	___ Exploring	___ Compromising

Scoring

T	S	C	J
_____	_____	_____	_____
Total numbers in these rows in the column above and write total on line above:	Total numbers in these rows in the column above and write total on line above:	Total numbers in these rows in the column above and write total on line above:	Total numbers in these rows in the column above and write total on line above:
2 3 4 5 7 8	1 3 6 7 8 9	2 3 4 5 8 9	1 3 6 7 8 9

*Developed by J. F. Veiga, University of Connecticut.

Copyright © 2008 by the Independent Electrical Contractors, Inc. All rights reserved.

The Leadership Style Inventory describes only your *perception* of your behavior. Get feedback from others to expand on this perception. Remember also that it describes how you behave as a leader in your *current* work environment. Styles are not fixed parts of our personality; rather, they represent how you have conditioned yourself.

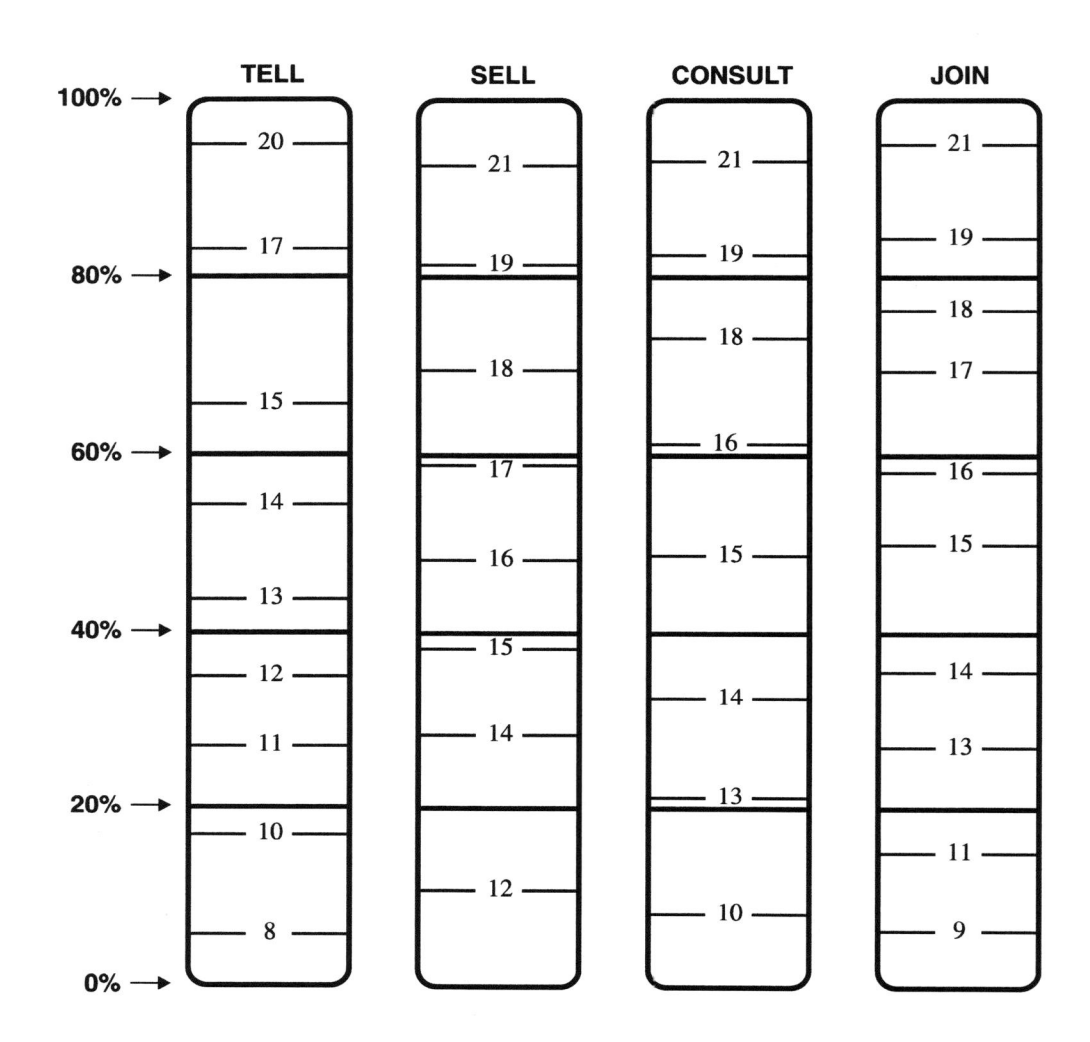

%	TELL	SELL	CONSULT	JOIN
100%	20	21	21	21
	17	19	19	19
80%				18
	15	18	18	17
60%		17	16	16
	14	16	15	15
40%	13	15	14	14
	12	14		13
20%	11		13	11
	10	12	10	9
0%	8			

**Leadership Style Profile
(Normative Data)**

The above chart can be developed into a profile of your leadership style. Shade in the area which corresponds to your score on each dimension. For example, if you scored 15 on the TELL scale, then shade the area up to the 15 under TELL on the above chart. The ruled-in percentile provides you a way of comparing yourself to others who have taken the inventory. The percentiles are keyed to style scores to indicate the number of people who scored below a particular score. For example, a score of 15 on the TELL style means you scored higher than almost 65 percent of the people tested.

Copyright © 2008 by the Independent Electrical Contractors, Inc. All rights reserved.

Building a Successful Team

Groups are common place in our society. Groups often develop when individuals come together in a particular time and space. Often in the workplace, groups are brought together to work on a particular project or accomplish a particular goal. However, groups may not necessarily be the most effective way to get a job done. By developing teams, work can become more productive and successful at accomplishing goals. The difference between a group and a team can at times appear small or very detailed. However, the real difference between the two can produce significantly different results. Some of the key differences between individuals and groups include:

Group

- Members work independently from each other;
- Members are directed what to do;
- Conflicts are often not dealt with or dealt with poorly;
- Distrust between members or supervisor is not uncommon;
- Members are afraid of making mistakes.

Team

- Members rely on each other to get the job done;
- There is a high level of trust;
- Good communication between members is practiced;
- Members are encouraged to learn and grow;
- Conflict is handled promptly and effectively.

A Successful Team

Working in a team that succeeds is especially satisfying. A group harmony develops when people who work closely together achieve a common goal, such as successful completion of a major project. This is called *esprit de corps*. The *Esprit de corps* that can develop among team members is very satisfying and cannot be equaled by those who work by themselves.

Whether we serve as members of an officially designated team or not, there are often opportunities to share in team rewards. For example, much of the class work that we do as an apprentices is done as part of a team. By pulling together and working well as a team, we can find ways to succeed in class, to learn more and to earn good grades in the process.

Whatever our situation, the more we can work closely with coworkers to set goals and achieve them, the more satisfying our work can be. Teamwork at any level has many rewards.

To be successful at teamwork:

- Act as a team member, even if we don't work in an official team. If we cooperate with others and help them find recognition, we can share their successes.
- Don't isolate ourselves to the extent that we are called loners. We can maintain our independence, but also participate as an enthusiastic member in a group endeavor.
- The team sees the satisfaction of the customer as center to the operation of the team.
- Responsibility, skill and control are shared.
- Individuals are expected to manage themselves and provide high quality work at all times. Everyone looks for mistakes or problems and works with the team to solve them.
- Everyone participates in problem solving.
- Accept and enjoy any form of team recognition. Find ways to celebrate your successes, even if there is no official recognition from management or instructors.

Building a successful team can be difficult to accomplish. Each team is different because each individual member is different. However, there are some things that you as the supervisor can do to help build a successful team. Keep in mind that building a team takes time and occurs after an extended process. While some groups come together more quickly as a team, there will certainly be ups and downs for any group forming a work team. The supervisor can set the tone by being willing to weather the inevitable ups and downs that can occur, while keeping the group focused on the job at hand. Some tools that the supervisor can use to build an effective work team while improving their leadership skills include the following:

- **A clear and motivating vision:** By providing your team with an motivating vision for their work, they will be more focused on the goals of the organization and project completion.
- **Listen to your workers:** While you don't always have to always implement their ideas, it's important that they feel you care about what they have to say.
- **Create an environment where team members trust each other:** Encourage working together, good communication is the rule, and never pit one worker against the other. The work setting should be viewed as a place that treats the employees fairly and with dignity.
- **Supporting employee self worth:** When your employees feel like they have the skills, resources, and support to do their job, they will be more motivated to do it well.
- **Clear policies and procedures:** People tend to work better and feel more confident when they are clear about the rules and expectations. However, the rules and expectations should not be so rigid that they can't be revised if needed.
- **Doing valuable and satisfying work:** Individuals want to feel that the work they do matters. Supervisors can really set the tone in this area by creating a work environment that supports this basic need.
- **Giving job feedback:** Although it might be difficult at times, giving feedback to team members can be a valuable to creating a satisfying work environment for the entire team.

- **Team involvement:** As the team leader, if you allow team members to participate in making decisions about their work, the members will have a highly level of "buy in" to the plan. Employees who believe in a plan tend require less supervision and have a higher job satisfaction.

Team Productivity

Increasing the productivity of employees is often a key business goal in companies across the United States. While this is an important goal for many businesses, it may not approached in a strategic way that allows workers to feel good about their work and grow in their skills. There are several strategies that supervisors can implement to increase team productivity and as a result, may also get increased employee job satisfaction. While these strategies may seem simple, or even not related to productivity, they can be the difference between average team productivity and great productivity. When working on team productivity, supervisors are encouraged to implement the following strategies:

- Work with team members to develop increased productivity goals. Perhaps you might want to consider decreasing work place injuries by a small percent in the next three months. Or increase the number of jobs completed in the next 30 days. The key points are that the goals are developed with employees and that they can be reached.

- Enlist the help of employees and upper management to remove the obstacles to team productivity. What is keeping the workers from working more efficiently and what, if anything, can be done to change it? For example, perhaps an inefficient way of making job assignments keep workers from beginning their work each day in a timely fashion. Why is that occurring and what can you as the supervisor, along with team members, do to improve the problem? Problem solving together can help build your team and increase team productivity as the same time.

- Team members should be aware of the goals of the team and company. Each member should know the steps to reaching the goals and should be committed to achieving them.

- Supervisors should work with each team member to grow and learn in their job. If a team member is having trouble completing their job duties, the supervisor (or perhaps another team member) should work with the individual to improve. Or perhaps the individual may need to be re-assigned to a position they are better trained to perform. Either way, it is important that each team member can feel good about their work and that they are rewarded appropriately.

- Keeping your team up to date on information regarding their job and the company can be critical in team productivity. If team members feel that key information is being withheld from them, productivity can take a quick nosedive. Rebounding from this can be very difficult and take a long time. Make a special effort to do regular updates with team members, while providing information about any major changes as they occur.

How to Use Rewards

While rewarding individual performance can be a very effective tool in maintaining and improving performance, rewarding team performance can be a nice complement to individual feedback. There can be three different main types of rewards. Each type is a large category of rewards that can be mixed and matched depending on the goals you are trying to achieve and the motivational factors of your employees. The three main types of rewards are as follows:

1. **Informal:** These types of rewards are more spur of the moment rewards. They are often easily implemented and require little or no pre-planning. They often have little or no cost associated with them. Included in this category are rewards such as an employer recognizing and congratulating an employee for good work, parking spot for employee of the month, or receiving a choice assignment for good performance.

2. **Awards for specific accomplishments:** This type of reward system includes rewards for meeting sales goals, giving teams decision making authority regarding work completion, and meeting safety goals. With this type of award, you need to be sure that the team members have a reasonable amount of control over their ability to achieve the goal, or there may be a reduction in team performance. For example, if you are giving rewards for a team meeting sales goals, and your product has a nationwide recall during that quarter, the team's ability to reach the goal is reduced.

3. **Formal:** This includes company developed formal reward programs. Examples of this type of reward include contests, incentive programs, and memberships to health clubs for meeting goals.

Rewards are most effective if the following criteria are met:

- The reward is viewed as desirable by the employee;
- It should be giving as soon after the accomplishment of the goal as possible;
- The amount of work required to meet the goal is not greater than the perceived value of the reward (the employee needs to feel the reward is worth working towards);
- The goal is viewed as achievable.

Team Communication

Well-trained employees have the confidence in their ability to contribute to the team effort. They understand why it is important to help support other members of the team. Research shows that the best team members are good communicators. These team members have learned to give clear instructions, stay responsive to questions and suggestions, and keep everyone on the team informed. Research also confirms that good communication leads to:

- Improved productivity
- Better problem solving
- A reduction in grievances
- Ideas for improvement in methodology
- Improved working relationships
- Greater job satisfaction

How to Use Team Communication

Many of the skills needed for effective team communication are the same ones you use when talking with just one other person, but they are more difficult to practice in a group situation. The following are tips for effective team communication:

1. **Be inclusive.** As you speak, talk to everyone. Move your eyes around the room. Don't just focus on one person. Seek input from everyone.

2. **Discourage dominance.** Don't allow one person to dominate the discussion.

3. **Be supportive.** Give positive recognition.

4. **Be sensitive to your teammate.** Pay attention to their behavior. Are they making eye contact? Are they acting differently that they were yesterday? Trying to understand them and how they are feeling will make you a better communicator.

5. **Ask for feedback.** One way to make sure that your message is getting through is to ask the other person.

7. **Keep emotions at a manageable level.** If a situation gets a little tense, a 10-minute break might be appropriate. You might also ask questions that lead the discussion in a less emotional direction for a while.

8. **Invite disagreement.** We often learn more through disagreement than through agreement. Disagreement can be productive.

8. **Be aware of how each member participates and responds.** Try to draw out a person who is sitting back saying nothing—but do it in a way that doesn't make him feel uncomfortable. If you see someone is becoming upset, intervene to protect his/her feelings.

9. **Use repetition.** Don't repeat so often that you are annoying. But ask the person to acknowledge your message and repeat it until they do understand your point.

Barriers to Effective Communication

There are several common **barriers** to better communications according to Harvey A. Robbins in his book, *How to Speak and Listen Effectively*. These barriers may also lead to conflict within the group or between groups.

- **Hearing what you expect to hear.** As human beings, we tend to listen for only those things that we expect to hear and we screen out everything else. That is why different people with different expectations hear different things.

- **Evaluating the source.** Human beings tend to evaluate the source of the information they receive to determine its value. This is why you tend to believe in and value what the President of the United States tells you if you voted for him.

- **Having different perceptions.** If you are upset with the way a co-worker communicates with you, you probably assume the co-worker wanted you to be upset. In reality, the co-worker was probably just communicating in a way that was comfortable for him/her.

- **Ignoring non-verbal communications.** Not all non-verbal signals are accurately interpreted. For example, most books on non-verbal attitudes will tell you that people who cross their arms and legs while listening are "tuning out". Not necessarily so! Many times the listener is displaying this behavior in order to concentrate better.

- **Being distracted by noise.** Usually it is more difficult to concentrate and communicate effectively when there's a lot of noise around us. To be able to better send and receive messages, it's preferable to find a quiet place to conduct business.

Resolving Team Conflict

There are five possible options for resolving conflicts:

- Withdrawing means that the project manager retreats from the disagreement. This is an option if the conflict is petty or of little impact on the project.

- Smoothing is used to emphasize areas of agreement to help minimize or avoid areas of disagreement. This is the preferred method when people can identify areas of disagreement and the conflict is relatively unimportant.

- Compromising involves creating a negotiated solution that brings some source of satisfaction to each party in the conflict. Compromises are best made after each side has had time to cool down. The best compromise makes each party feel as though he won.

- Forcing is used when the boss exerts his or her position of power to resolve a conflict. This is usually done at the expense of someone else and is not recommended unless all other methods failed to resolve the conflict.
- Confronting is not quite as strong as forcing, but it is the most common form of conflict resolution. The goal of a confrontation is to get people to face their conflicts directly, thereby resolving the problem by working through the issues in the spirit of problem solving.

Building Consensus/Team Problem Solving

The question that often comes up when groups try to solve problems is how do we do this quickly and effectively. There are several phases of communication that we need to move through to do consensus problem solving.

- We should generate ideas by discussion, using open questions and brainstorming creatively
- We should record all the input somewhere so that everyone can see. Get everyone to understand others' ideas by reviewing what has been recorded. Then, work together to evaluate all the input. The challenge is to keep everyone focused on finding the solution and not wandering off on tangents.
- Be open-minded, listen fully and respect others' views. Acknowledge others contributions and provide feedback on their ideas. Propose solutions to differences and be willing to negotiate. Identify areas of agreement and work from there to seek consensus.
- We should work out the best solution by summarizing, eliminating and narrowing. When we have narrowed down the possibilities, then we can quickly rank the solutions and prioritize what we think are the top solutions to the problem.

Project Management

During most of your working life you will be working on projects; however, you may not think that you need to know about project management. However, everyone on the job needs to understand project management and how to implement it on a day to day basis. The very idea of project management may sound too big, too complicated and too time consuming. But some day, even some day very soon, you may need to know about how to manage a project.

This section of the handbook is intended to give you an overview of the important components of project management and how to use them. The purpose of these mini-lessons is to help you if and/or when you are asked to help manage a project at work. In addition to these mini-lessons, you should pay attention to how project management is done at your company. For instance, how does your company chart the progress of a project? How do you do a budget? How do you finish off a project? This handbook will outline the key areas of project management.

Introduction to Project Management

Project management is a system for managing tasks, resources and costs effectively. Managing projects means keeping scope, schedule and resources in balance. The scope of the project is the number of tasks required to get the project completed. The schedule refers to how you keep track of the time and sequence of each task in order to complete the project on time. Resources are the people, equipment, and materials need to do the project. Since people, equipment, and materials cost money, you need to keep track of these to make sure that the project is within budget.

Good project management helps you succeed. Everyone on the team will benefit from improved coordination and use of resources. Project management helps everyone have a sense that everything is under control.

There is no single way to manage a project. But every project, generally, has four parts:

1. Defining the Project
2. Planning the Project
3. Implementing and Tracking the Project
4. Closing the Project.

Defining a Project and Tasks

What are tasks? Why break your project into tasks? A project task is a related set of work steps that help to get a project done. A list of tasks can be short and simple or long and complex, depending on the size and goal of the project. There are several reasons for organizing your project into smaller work units called tasks:

- By breaking work into tasks, you can put the in work in your project in the most logical sequence for completion. This can help you determine the best schedule for your project.
- Tracking the progress of tasks is a way to assess and control the work done on a project.
- By breaking the work into tasks, you can determine the skills you need to complete the work on a project and you can determine how many people will be required to do the work. Tasks allow you to communicate the work that needs to be done to other team members. A task list can be used to negotiate changes in work sequences during the project.
- Breaking the work into tasks ensures that all work sequences are identified and understood.

There are several characteristics of good task that is may help to keep in mind.

- A task should state clearly what needs to be done.
- All work within a task should occur in sequence without gaps for other tasks in between. For instance, "Framing and wiring the library" is two tasks, not one.
- Most steps in a task should use the same team members. If it requires a different set of skills and a different team member, it is probably a different task.
- A task that involves multiple people usually goes smoothly if all the people working on the task are at the job site and within walking distance of each other.
- You can use the 8/80 rule and the reporting period rule to scope the size of the task. The 8/80 rule means that you keep your tasks between eight hours (one work day) and 80 hours (10 work days). The reporting period rule states that no task should be longer than the standard reporting period. Thus, if you make weekly reports, no task should be longer than a week. This helps you keep tasks on the same schedule as your reporting cycle. (Both these rules are guidelines, not requirements.)

Planning

Planning is an important part of project management. Planning allows you to establish what is required to successfully complete the project. Planning helps you make sure that quality standards are met, timelines are adhered to and costs are kept within budget. What follows are some steps for planning:

- Establish the project goal;
- Choose the basic strategy for achieving that goal;
- Breakdown the project into steps or tasks;
- Determine the performance standards for each step or task;
- Determine how much time is required to complete each task;
- Determine the proper sequence for completing each task;
- Determine the cost of each task;
- Decide who on your team will need to do each task;
- Develop a work plan that outlines all the information from the steps above.

Implementing a Project

Every project has three primary goals:

1. To create something;

2. To complete it within a specific budgetary framework;

3. To finish within an agreed upon schedule.

Beyond these goals are the other goals that must be specified that actually define the project. To distinguish primary goals from other project goals, the other goals are sometimes referred to as objectives. But goals and objectives are usually different words for the same thing. No matter what they are called, they should meet the following criteria:

- Goals must be specific. Your goals should be so clear that another project manager should be able to take over for you, if need be. Also, all of the project team should be able to read and understand your goals. Confused, surprising or conflicting responses mean that you have more work to do.

- Goals must be realistic. Your goals must be within the realm of possibility. For instance, if similar projects have taken 10 weeks to finish and you set this goal at 2 weeks. That may not be possible, even with crews working 24/7.

- Goals must have a time in which they must be accomplished. Projects must have a definite finish date or they may never be completed. Projects with no ending date never finish. Projects with unrealistically short dates blow up like a fuse box hit by electricity.

- Goals must be measurable. You must be able to measure your success at meeting your goals. These results are called deliverables—the results of the project. Quality is a big part of these deliverables. Deliverables, like projects, are evaluated not only by the fact that something is accomplished but also by the quality of the work as well.

- Goals must be agreed upon. You, your boss, and your customer must agree on the goals before you take any further steps toward planning the project. If you don't reach consensus, there's no point in beginning the project; it is doomed from the start.

- Responsibility for achieving the goals must be identified. Although you as project manager bear the brunt of responsibility for the overall success of the project, others may be responsible for pieces of goals. The people responsible for the goals must be identified and be willing to accept responsibility before the project proceeds. It is best to have the sign-off of all the major players before you begin the project.

Planning for quality requires attention to detail. The goal of quality planning is to assure that the finished project will do what it is supposed to do. A quality plan also establishes criteria by which the project will be evaluated when it is finished. In planning for quality, you need to include the quality of materials to be used, the performance standard of the materials used, and the means of verifying quality, such as testing and inspection. Two techniques—work breakdown structure (WBS) and project specifications—facilitate planning for quality.

A work breakdown structure (WBS) is the starting place for planning for cost and time, as well as quality. It is a technique that divides a project into steps to be completed in a sequence. Because all elements required to complete the project are identified, a work breakdown structure reduces the

Project Management and Team Building Handbook—19

chances of neglecting or overlooking an essential step. A work breakdown structure usually has two or three levels of detail, although more complex projects may require more detail. As you construct a work breakdown structure, keep in mind that the goal is to identify distinct units of work or tasks that will help you move the project toward completion. The work breakdown structure can be used to assign project task responsibilities, to build a budget, and to look for tasks that require excessive amounts of funds. You also can rough out the schedule and time required, but for many projects you may want to use more targeted charting, like Gantt or PERT charts.

A way to judge the lowest-level task you should accommodate in your work breakdown structure is to consider budget or time criteria. The lowest-level task should require at least .25 to 2 percent of the total budget or project time. You can also use segments, such as half or full days, to define the smallest task levels. These approaches help establish general guidelines for the lowest level tasks in your project.

There is no magic formula for organizing the work breakdown structure. You may do it based on the category of the task. For instance, in projecting for constructing an office, there may be a task called "Finishing" that may include breakdowns of painting, installing floor covering, installing electrical fixtures and cleanup. You may do the WBS based on technological disciplines. For instance, there may be a task called "Electrical" that may include conduit, wiring, fixtures and testing. You may do the WBS based on organizational structure. For instance, some firms break out tasks between the electrical and telecommunication sides of the company. You may do the WBS based on physical location. If you are doing work at more than one location, build the WBS based on geographical locations instead of people. Lastly, you may do the WBS based on systems and subsystems.

Don't forget to put project management tasks in the WBS. These tasks take time, too. These tasks might be grouped under a task called "Managing the project," which includes time for developing reports, holding meetings, ordering materials and other items. Also, when maintaining a WBS for large projects, you should have a version control system. This means that each version of the WBS should be date- and time-stamped. That way older plans can be stored in the file cabinet and everyone works from the most current version. Older version can be saved so they can be used to track project history and as a learning tool to evaluate what went wrong.

Spending time up front to develop a good plan can be one of the most useful things that you can do for project management. According to Steve McConnell in Rapid Development, "If a defect caused by incorrect requirements is fixed in the construction or maintenance phase, it can cost 50 to 200 times as much to fix as it would have in the requirements phase. Each hour spent on quality assurance activities such as design reviews saves three to 10 hours on downstream costs."

From the work breakdown structure, specifications can be written for each step of the project. Specifications include everything that is necessary to meet the projects quality dimension, including material to be used, standards to be met, tests to be performed, and so on. Extreme care should be used in writing specifications, because they become the controlling factor in meeting project performance standards. They also directly affect budget and schedule. Some examples of project specifications include:

- All cables outside of the telecommunication room will be supported utilizing approved cable supports, a minimum of 48 inches apart.
- No horizontal cable shall exceed 295 feet in length.
- Patch Panel icons will reflect the colored icons on the jacks at the workstation.

- Cable testing shall be completed according to the most current version of EIA/TIA 568 and the associated TS bulletins.

Now take the current project you are working on writing specifications for at least one step of the project:

Step:

Specifications:

Planning Time: The idea when planning the amount of time for each project task is to determine the shortest time necessary to complete each task or step. Begin with the work breakdown structure and determine the time necessary to complete each step. Next, determine in what sequence the steps must be completed, and what steps can be under way at the same time. From this analysis, you will determine the three most significant time elements: (1) The duration of each step, (2) The earliest time at which each step may be started, (3) the latest time by which a step must be started.

Planning the time for each task needs to be done by people who have experience with the activities designated in each task. The experienced project manager can look at a task, understand what is required, and provide a reasonable realistic "guesstimate" of the time required. You will get to that point someday if you start paying attention to the time it takes to accomplish things now. By paying attention, you will build up an understanding of the time involved. However, when it is your turn to build a schedule of work, if you are unsure of how long it takes to do something, you will need to rely on someone who does have the required experience.

Most project managers find it realistic to estimate time intervals as a range rather than as a precise amount. Because this is like trying to predict the future, you can only guess. But there are better ways to guess than others. When estimating the duration of tasks, there are five options that you can use to make the estimates as good as possible:

- Ask the people who will actually do the work, but have them estimate the work they'll be doing- not the durations. You will need to add extra time based on their workload, their other commitments and the experience with similar tasks.

- Get an objective expert's opinion (for example, someone who has worked on a similar project).

- Find a similar task in a completed project to see how long it took to get it done.

- If you have time and the task lends itself to this, perform a test session of the task to see how long it really takes using your resources.

- Make your best educated guess. This is the last resort when you are under pressure, so make sure the guess is based on as much experience as possible.

Some project managers may want to start with the budget first, but that is usually not a good idea. Because time is money, the schedule will affect the final budget in many ways. That's why it is a good idea to talk about schedules first. It is best to try to determine how much time is really required to complete a project before you let your money (or lack of it) get in the way of your thinking.

Project Management and Team Building Handbook–21

Take some time to list some of the tasks from the project you are currently working on (or have just completed) and write down estimates of the time that you think it will take to complete each task.

Project Costs: A good plan is doing research on sources of supplies and materials to assure that estimated costs are realistic. If you overestimate costs you may lose the job before you begin because your rates are not competitive. Some inaccuracies in budget are inevitable, but they should not be the consequence of insufficient attention while drafting your plan. The goal is to be as realistic as possible.

You cannot estimate the cost of your project until you know how long it will take, since labor is typically the highest cost item. Therefore, you should use your tasks and project schedule as the starting point for developing your project budget. Typical budget areas include:

- Labor. The wages paid to all staff working on the project for the time spent on the project.
- Equipment. The cost of purchasing or renting equipment for the project.
- Materials. The cost of items purchased for use in the project, such as cable, splicing kits, etc.
- Supplies. The cost of tools, office supplies and other general items needed by the project.
- Overhead. The cost of fringe benefits, usually a percentage of labor costs.
- General and Administrative. Cost of management and support services, e.g. payroll or purchasing.
- Profit. The reward to the firm for successfully completing the project, usually a percent of the cost.

With the cost components identified and the project broken down into steps, you can create a worksheet or spreadsheet to total the costs.

There are areas of potential budget problems that you need to be aware of. First, costs may go up between the time that you make up the budget and the time that you start the project. If you fail to get firm price commitments from suppliers and subcontracts, cost increases can cause problems for your budget. Secondly, poorly prepared project plans can lead to incomplete budgets. It is important to consider every task involved to get the job done, every resource needed to accomplish that task and budget amounts for everyone of these resources. Lastly, you need to make sure that every resource is available. For people, this means knowing their work availability, vacation schedule and commitment to other projects. For other resources, this means checking that all items will be available when you need them. If people or things are not available when you need them, it may cost you money that you didn't budget for.

Once a project is underway, your key responsibility is to keep things going—on time and on budget. Getting a handle on your project isn't really hard. You have a plan and now is the time to use it. Your plan is the main tool for maintaining control of the project through its lifetime. This requires careful management of both the plan and its resources, including people. You will need to consistently monitor and update the plan. You should use the project specifications to monitor the quality of the project. You may need to update as conditions change so that it reflects the current status of the project, in terms of budget constraints, schedule changes or product modifications. You must monitor the progress of the project against the plan on a regular basis. Project managers need to compare the time, cost and performance of the project to the budget, schedule and tasks defined in the approved project plan. This should not be done in a haphazard way, but done on a regular basis. Your bosses will need to know about any significant departures from the budget or the schedule, otherwise these departures could affect the success of the project.

Remember that quality communication is a key to good project management. Every person involved in your project requires ongoing communication at various levels of detail. Not everyone needs to know everything, but everyone needs some information to help him or her do their part of the project. Some communication will be formal, some will be informal. The objective of communication is to keep people informed, on track, and involved in the project. It is up to you to get involved with all aspects of the project so that you keep track of what is going on in the project and contribute to the work being done.

Tracking Progress

A helpful technique for controlling a project is to invest some time thinking through what is likely to go wrong in terms of quality, cost and timeliness. Then identify when and how you will know that something is wrong and what you will do to correct the problem if it occurs. This will help minimize the likelihood that you will be caught by surprise, as well as save time in responding to the problem. A chart like the one below, called a Control Point Identification Chart, is an easy way to summarize this information.

Control Element	What is likely to go wrong?	How and when will I know?	What will I do about it?
Quality	e.g. Workmanship will be less than desired	e.g. Upon personal inspection of each stage of the project	e.g. Have substandard work redone.
Cost			
Timeliness			

A helpful tool for project management is a project control chart. This chart uses budget and schedule plans to give a quick status report of the project. It compares actual to planned, calculates a variance on task that is completed and tallies a cumulative variance for the whole project. To prepare a project control chart, list all the tasks for the project on a spreadsheet. Then use the schedule to list the time planned to complete each step and use the budget to list the expected cost of each task. As each task is completed, record its actual time and actual cost. Calculate the difference between the two. This is known as variance. Total all the variance to give you a quantity for how much the project is different from what was planned.

Another kind of project control chart is called a milestone chart. A milestone chart presents a broad-brush picture of a project's schedule and control dates. It lists those key events that require approval before the project can proceed or that need to be reported in communication with the bosses and/or the project team. If this is done correctly, a project will not have many milestones. Because of this lack of detail, a milestone chart is not very helpful during the planning phase when more information is needed. However, it is particularly useful in the implementation phase because it provides a concise summary of the progress that has been made. A milestone simply lists the task in the first column, the scheduled completion date in the second column and the actual completion date in the third column.

Project Management and Team Building Handbook–23

Budget control charts are generally of two varieties. One lists the project tasks with the actual costs compared to budgeted costs. It is similar to the project control charts, described above, and can be generated by hand or by computer. The other kind of budget control chart is a graph of budgeted costs compared to the actual costs. Either bar or line graphs may be used. Bar graphs usually relate budgeted and actual costs by project task, while line graphs usually relate planned cumulative project costs to actual costs over time. Another helpful approach to budget control is to compare the percentage of budget spent to the percentage of the project completed. The data can be compared by making a list or a graph. While the percentage of budget spent is a precise figure, the percentage of the project completed should be your best estimate of the project progress.

Risk Management: You can reduce the risks on your project through risk management. Some people even refer to project management as the practice of risk management. In basic risk management, you plan for the possibility that a problem will occur by estimating the probability that the problem will arise during the project, evaluating the impact if the problem does arise, and preparing solutions in advance. All risk management starts with identifying the risks. Identifying risks takes careful analysis. Assume that anything can go wrong. Learn from past projects. The failures of the past are often the best source of riskcontrol information. It also helps to look at deliverables from different points of view including those of the workers, subcontractors, vendors, suppliers, management and customers. You should also examine critical relationships and what can happen if someone quits or doesn't deliver.

To analyze the probability that a risk will occur and the potential impact of the risk, assign a number on a scale of 1 (lowest probability or impact) to 10 (highest probability or impact). Assign a number for both importance and for probability. Use this to help you determine the overall severity or importance of the risk, by multiplying the probability number by the impact number. This is a measure of severity. Usually an item with a number more than 40 requires careful analysis and monitoring.

Develop a response plan for the risks. You have four basic options for dealing with risks on your list:

1. **Accept the risk.** This means you do nothing special at this point. If something happens, you deal with it. This is appropriate when the consequences for the risk are cheaper than a program to eliminate or reduce the risk.

2. **Avoid the risk.** This means you delete part of the project that contains the risk or break the project into smaller tasks that reduce the risk overall. However, sometimes you will want to take on more risk to earn more return.

3. **Monitor the risk and develop a contingency plan in case it appears that the risk will happen.** Developing contingency plans for key risks is one of the most important aspects of risk management. These contingencies are alternative plans and strategies to put in place when necessary. The concept is that you will do better thinking about things when there is not a crisis on your hands.

4. **Transfer the risk.** Insurance is the most obvious way of transferring risk. Another way of transferring risk is subcontracting. Lastly, a fixed price contract with a vendor is another way of reducing budgetary risk in a project.

Remember that risk management and response planning are ongoing processes. Risk must be regularly evaluated. Every project encounters problems neither planned for nor desired, and they will require some type of action before the project can continue.

The heart of project management is monitoring work in progress. It is your way to know "what is going on" and how "what is going on" compares to what was planned for. With effective monitoring, you will know if and when corrective action is required. Since the status is constantly changing, you'll need to monitor the project and compare to the plan in some way every day. Reports are good tools for combining information, but informal discussions often reveal a more accurate picture of the project.

Following are ways to keep abreast of project progress:

- Inspection is probably the most common way to monitor project progress. Inspection is an effective way to see whether project specifications are being met, as well as whether there are unnecessary waste or unsafe work practices. Ask questions and listen to explanations.

- Interim progress reviews are communications between the project manager and those responsible for various steps of a project. Progress reports typically occur on a fixed time schedule or at the completion of each task. Three topics are usually covered during these reviews: (1) Review of the progress compared to the plan, (2) Review of problems encountered and how they were handled, (3) review of anticipated problems with proposed plans for handling them.

- Testing is another way to verify project quality. Certain tests are usually written into the specifications to confirm that the desired quality is achieved. Typical tests for cable installation include testing for continuity, opens or short and according to EIA/TIA standards.

Auditing can be done during the course of the project or at its conclusion. Common areas for audit are financial, maintenance procedures, purchasing practices, and safety practices. Auditors should be expert in the area under review.

As a project progresses and you monitor performance, there will be times when the project does not measure up to the plan. This calls for corrective action. However, don't be too quick to take action. Some deficiencies are self-correcting. It is unrealistic to expect steady and consistent progress day after day. Sometimes you fall behind and sometimes you'll be ahead. The successful project manager is able to maintain a perspective that allows honest, objective evaluation of each issue as it comes up. The key is to monitor things so that nothing gets out of hand.

When quality is not according to specification, the customary action is to do it over, according to plan. However, if the work or material exceeds specifications, you or your client may choose to accept it.

When the project begins to fall behind schedule, there are three alternatives that may correct the problem. The first is to examine the work remaining to be done and decide whether the lost time can be recovered in the next tasks. If this is not feasible, your company may want to offer an incentive for on-time completion of the project. The incentive could be justified if you compare the cost to potential losses due to late completion. Sometimes a supplier can deliver a partial order to keep your project on schedule and complete the delivery later. Finally, consider bringing more resources. This too will cost more, but may offset further losses from delayed completion.

When a project begins to exceed budget, consider the work remaining and whether or not cost overruns can be recouped on work yet to be completed. If this isn't practical, consider narrowing the project scope or renegotiate with the client. Perhaps nonessential elements of the project can be eliminated, thereby reducing costs or saving time. When something is not available or is more expensive than budgeted, substituting a comparable item may solve your problem. Or you may need to look for other suppliers that can deliver within your budget and schedule. Sometimes demanding that people do what they agreed to do gets the desired results. You may have to appeal to higher management for backing and support. The goal of project management is to obtain client acceptance of the project's end result. This means that the client agrees that the quality specifications were met. In order to have the acceptance stage go smoothly, the client and the project manager must have well-documented criteria for judging performance. These should be objective, measurable criteria that are not subject to interpretation. There should be no room for doubt or ambiguity, although this is often difficult to achieve. It is also important to be clear what the project output is expected to accomplish.

Closing a Project

The project may or may not be complete when results are delivered to the client. Often there are documentation requirements and a final report that still need to be provided. Finishing a project takes time and deliberate effort on your part. The final step of any project should be an evaluation review. This will look back over the project to see what was learned that will contribute to the success of future projects. This review is often best done as a team with those working on the project.

The following is a project completion checklist:
- ✔ Test project output to see that it works
- ✔ Write operations manual (if needed)
- ✔ Complete final drawings
- ✔ Train client's personnel to operate the project output
- ✔ Reassign project team
- ✔ Dispose of surplus equipment, material and supplies
- ✔ Summarize major problems encountered and their solutions
- ✔ Summarize lessons learned
- ✔ Write performance evaluation reports
- ✔ Provide feedback on performance to all project staff
- ✔ Complete final audit
- ✔ Write final report
- ✔ Conduct project review with upper management
- ✔ Declare project complete

Project Management and Team Building

Becoming a Supervisor

- Use good communication skills
- Be objective
- Be "real"
- Motivate your workers
- Delegate
- Be respectful of differences
- Let go of your old job

Your Leadership Style

- Tell
- Sell
- Consult
- Join

Leadership Inventory Style

- Page 8 in Team Building and Supervision Handbook
- Review Instructions
- Complete inventory

Building a Successful Team
Group vs. Team

Group
- Members are independent
- Members are directed
- Conflicts not dealt with
- Distrust
- Afraid of making mistakes

Team
- Members rely on each other
- Members trust each other
- Good communication
- Growth is encouraged
- Conflict is dealt with effectively

Team Productivity

- Set productivity goals
- Remove obstacles
- All should be aware of team and company goals
- Work with team members to learn and grow
- Keep your team up to date on latest information

Using Rewards

- Informal
- Awards for Specific Accomplishments
- Formal

Team Communication

- Be inclusive
- Discourage dominance
- Be supportive and sensitive
- Ask for feedback
- Keep emotions down
- Invite disagreement
- Be aware of individual differences
- Use repetition

Resolving Conflict

- Withdrawing
- Smoothing
- Compromising
- Forcing
- Confronting

Building Consensus/ Problem Solving

- Clearly idetify the problem
- Generate ideas and possible solutions
- Record ideas
- Be open-minded and respectful
- Find best solution

Project Management

- Introduction
- Define the project & tasks
- Plan
- Implement the project
- Track progress
- Close the project

Team Building Action Plan

- What are three of my current strengths that I can bring to building a successful team?
- What steps can I take to build on those strengths? List three concrete steps you can take.
- What are three skills that I want to work on related to team building?
- What steps can I take to gain the team building skills I need? List at least three concrete steps you can take.

Lesson 416 – Project Management and Team Building Lab Checklist

Student Name:_____ Date: _____

Skill The Student:	Level 1 unskilled	2 average	3 excellent	Initials
Works with their team towards a common goal.				
Participates actively in project development and presentation.				
Assists with conflict resolution and problem solving.				
Uses good communication skills.				
Recognizes accomplishments of members of their team.				
Gives appropriate feedback to other team members				
Motivates self and others.				
Demonstrates project management skills.				
Can be counted on to fulfill commitments.				

Tools and Equipment The Student correctly used the following:		YES	NO
	Leadership Inventory		
	Project Management and Teambuilding		
	Professionalism lesson		

Lab Checklist – Lesson 416 – Project Management and Team Building – Page 1

Lesson 416 – Project Management and Team Building Lab Checklist

Student Name:_____ Date: _____

Inspection			
The Student's team and individual work passed the following inspections:			
	Gave a succinct and coherent presentation.		
	The presentation addressed safety issues and codes and standards (as applicable).		
	Used an activity or exercise as part of the presentation.		
	Developed and used a set of review questions to help other students focus on important points and topics.		

Overall Lab Performance Score _____

Checker's Signature_____

Date:_____

Overall Lab Performance Score _____

Checker's Signature_____

Date: _____

Comments:

_____

_____

_____

_____

_____

Lab Checklist – Lesson 416 – Project Management and Team Building – Page 2

Semester Review (Student Manual) Lesson 417-1

IEC
PRIDE
NATIONAL

Electrical Curriculum

Year Four
Student Manual

Lesson 417 – Semester Review

Purpose

To review Lessons 404 through 416. Prepare students to take the Semester Exam.

Objectives

By the end of this lesson, you should be able to:

404-1	Utilize the basic printed circuit (PC) Board and its main components
404-2	Analyze semiconductor theory and its relation to semiconductor devices
404-3	Categorize N-type and P-type material
404-4	Classify rectification systems
404-5	Differentiate among various types of diodes
404-6	Explain the theory, operation, and use of various other solid state devices
405-1	Discuss the operations and functions of various electromechanical relays
405-2	Demonstrate proper contact arrangement and terminology
405-3	Describe the operation and functions of solid state relays
405-4	Demonstrate the selection and installation of the proper relay for an application
405-5	Identify the advantages and disadvantages of different types of relays
406-1	Utilize solid state control devices
406-2	Operate electromechanical and solids state relays
407-1	Evaluate when and how to use a photoelectric control in a control circuit
407-2	Explain how the different types of photoelectric controls function
407-3	Discriminate among the different types of proximity switches and how they function
407-4	Explain the Hall effect
408-1	Explain the uses of the programmable controller
408-2	Relate the functions of the parts of a programmable controller
408-3	Identify I/Os and write a simple program for a programmable controller
408-4	Categorize programmable controller applications
408-5	Identify the advantages of using multiplexing for specific applications

Copyright © 2008 by the Independent Electrical Contractors, Inc. All rights reserved.

409-1   Install and connect photoelectric and proximity controls
409-2   Hook up and use a programmable controller

411-1   Determine the reasons for reduced voltage starting in AC and DC motors
411-2   Explain how a primary resistor starter works
411-3   Explain how an autotransformer starter works
411-4   Explain how a part-winding starter works
411-5   Explain how a wye-delta starter works
411-6   Explain how a solid-state starter works

412-1   Distinguish among various motor braking techniques
412-2   Explain multispeed motors and their connections
412-3   Utilize various ways to control the speed of a motor

413-1   Practice using various reduced voltage starters
413-2   Employ the brakes and control the speed of various motors

414-1   Specify preventative maintenance techniques and programs
414-2   Demonstrate proper troubleshooting techniques
414-3   Evaluate why motors fail

415-1   Troubleshoot motor control systems

416-1   Identify your personal leadership style
416-2   Develop project management skills
416-3   Develop skills in communicating effectively with others in a team setting
416-4   Develop skills in building successful work teams

Content

Knowledge and Skills

This lesson is designed to instruct you in the following:

- Different types of solid state devices.
- Electromechanical and solid state relays.
- Photoelectric and proximity controls.
- Programmable controllers.
- Reduced voltage starters.
- Acceleration and deceleration methods for motors.
- Preventative maintenance and troubleshooting methods.
- Operating characteristics of TRIACs, DIACs, and transistors.

Semester Review (Student Manual) Lesson 417-3

Professionalism

- Work as a life-long learner to keep abreast of constant changes in the industry
- Ability to plan and organize work materials
- Ability to manage tasks, resources and costs effectively
- Ability to set goals and see them through to completion
- Recognize and promote project quality
- Ability to use project specifications

Project Specifications

From the work breakdown structure, specifications can be written for each step of the project. Specifications include everything that is necessary to meet the project's quality dimension, including material to be used, standards to be met, tests to be performed, and so on. Specifications are the controlling factor in meeting project performance standards. They also directly affect budget and schedule.

Using a current project on which you are working try writing specifications for at least one step of the project:

Step: _____

Specifications: _____
_____

Relevant Tools and Equipment

- Hand tools
- Soldering iron and solder
- Alligator clips leads
- Digital or analog Volt-Ohm meter
- Oscilloscope
- Continuity meters and beepers
- Multimeters
- Wire jumpers
- Power sources
- Amprobe

Code References

- None

Copyright © 2008 by the Independent Electrical Contractors, Inc. All rights reserved.

Homework

1) Reading Assignment

- Review all your homework assignments for Lesson 404-416.

2) Key Terms and Formulas

Review all key terms and formulas for Lesson 404-416.

- $T = \dfrac{5252 \times HP}{rpm}$
- $Work = Distance \times Force$

- $HP = \dfrac{I \times E \times eff}{746}$
- $I = \dfrac{E}{R}$

- $HP = \dfrac{rpm \times T}{5252}$
- $I = \dfrac{E - C_{EMF}}{R}$

- $V/HZ = \dfrac{V}{HZ}$
- $S_{rpm} = \dfrac{120 \times F}{P}$

- $PD_m = \dfrac{PD_d \times N_d}{N_m}$
- $Power = \dfrac{Work}{Time}$

- $rpm = \dfrac{120 \times f}{N_p}$
- $V_u = \dfrac{V_d}{V_a} \times 100$

3) Practice Exercises

- None

Semester Review (Student Manual) Lesson 417-5

Study Tips

▶ There are a few hints to remember when you take tests like the Semester Exam.

   A. Be sure that you understand the instructions.

   B. Read all the questions first. Then answer the easiest questions first. This will help build your confidence and ensure that you get credit for the easy questions.

   C. Don't struggle with a question and waste your time. Go on and come back to it later.

   D. Make sure that you allow enough time to finish all the questions.

   E. If you are unsure of an answer, put down what you think is the best answer and come back to it later. Something in the rest of the test may trigger the correct answer for you.

▶ **A good review technique is to develop a Question and Answer (Q&A) Outline.** A Q&A Outline is a summary of an entire textbook or CD chapter or article that is constructed by identifying "questions" that the content "answers." The focus of the outline is on the content's main points. All textbook material can be regarded as answers to implied questions. Identifying these questions and linking them to answers helps in understanding and recalling the information. It also serves as a study manual for studying for tests.

▶ In general, the questions should all end in question marks, should not be able to be answered by "yes" or "no" answers, should cover all the main points, should be consecutively numbered, should have brief answers with page number locations.

▶ What are different kinds of questions in a Q&A outline?

  - **Recap questions** are questions whose answers summarize entire sections or subsections of text. They are typically constructed by converting headings into question form. They can make up all or most of the entire Q&A outline (if the text has enough headings).

  - **Reflection questions** are questions whose answers summarize paragraphs. They are typically constructed on the basis of the first sentence in a paragraph, or words in italics or boldface or items in numbered or bulleted lists.

  - **Reasoning questions** are questions whose answers are related to, but go beyond information in the text. They often look like short answer or essay questions on a test. They are constructed by asking for an explanation, application, comparison or evaluation of ideas in the text. The answers cannot be contained in the text.

Semester Final Exam (Student Manual) Lesson 418-1

IEC NATIONAL

Electrical Curriculum

Year Four
Student Manual

Lesson 418 – Semester Final Exam

Purpose

To evaluate your progress throughout the semester, you will take a final exam of approximately 100 questions from the test banks for lessons 404-416.

Series Power Analyzers

...Power Quality Analyzer is a versatile and easy-to-use tool that will give you the ... you need for power quality analysis. The 800 Series Power Analyzer measures power ...oth single-and three-phase electrical systems. Instantaneous viewing of current, ...ver and harmonics including tables, bar graphs and even individual waveforms, make ... interpretation easy. With a real-time clock and 1MB of memory, the Power Analyzer ...that can be downloaded to a personal computer for in-depth analysis using ...™ software.

...ng a professional needs for monitoring and analysis in a single unit: rugged, versatile ... use. Models available with 3 or 4 current channels to meet your needs.

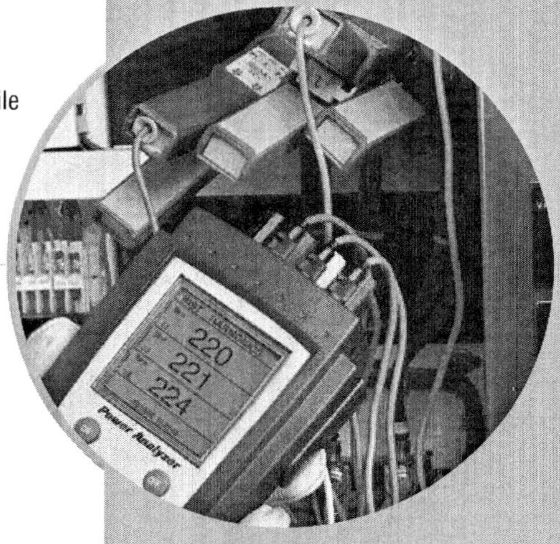

...se Power Analyzer

Features

- ...e RMS
- ...gle- or three-phase measurements
- ...a logging (1MB memory)
- ...rgy and harmonics program ...uded
- ...r voltage channels
- ...r current channels*
- ...erVision™ Software
- ...ntweight, portable design

- Easy-to-use push-button operation
- Backlit display
- Customized data collection
- 2-year warranty
- Intelligent battery charging system
- Easy set-up
- Interactive demo CD

...del 805 equipped with three channels for ...ent

- **Measures power quality in commercial and industrial electrical power systems**
- **True RMS readings for accurate testing**
- **1 MB of memory stores data for in-depth analysis through PowerVision™ Software**
- **Rugged, compact design**

...nics Measurements

- ... harmonic distortion (% THD)
- ...onic factorization (FFT) to 51st ...onic

Power & Energy Measurements

- Kilowatts (KW)
- Volt-amperes (VA)
- Inductive reactive power (kvarL)
- Capacitive reactive power (kvarC)
- Power factor (PF)
- Frequency
- Kilowatt hours (kWh)
- Reactive power per hour (kvarhL, kvarhC)

PowerVision™ Software

...oftware provides intuitive analysis of data and automatic ...seful tables and graphs. Files are easily downloaded from any

...rovides multiple ways to view basic and detailed ...owing you to customize readouts according to your trouble ... Compare up to 12 variables simultaneously. Easily scroll ...d and collapse data over time. Compare data from multiple ...fore and after studies. Data can be exported to text files. ...hs and tables can be pasted into Excel or Word reports. ...echnical support people are always a phone call away.

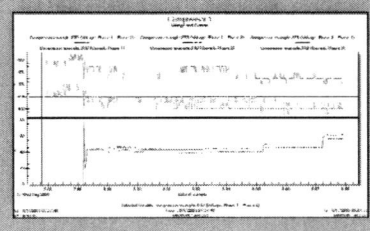

Current and Voltage 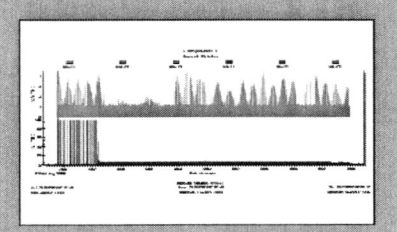 | Harmonic Distortion

Using Digital Multimeters to Diagnose Power Quality—Lab (Student Manual) Lesson 419-1

IEC NATIONAL PRIDE

Electrical Curriculum

**Year Four
Student Manual**

Lesson 419 – Using Digital Multimeters to Diagnose Power Quality—Lab

Purpose

This lesson will focus on the effects and problems that result from poor power quality. The lesson will build on your prior knowledge of digital multimeter use. You will also apply what you have learned in previous lessons about AC voltage and current measurements, harmonics, and 3-phase circuits. This lesson will conclude with a conceptual overview of power conditioning and protection.

Objectives

By the end of this lesson, you should be able to:

419-1   Describe major power quality problems and sources
419-2   Discuss the basic concepts and procedures for power quality measurement
419-3   Discuss the effects of poor power quality on electrical equipment and systems
419-4   Discuss indicators of power quality problems and the equipment and methods used to resolve those problems

Content

Knowledge and Skills

- Identify general sources of power quality problems that can effect load operation
- Describe how voltage changes can be a power quality problem
  ➢ Power interruption
  ➢ Voltage fluctuations
  ➢ Voltage sags and swells
  ➢ Undervoltage and Overvoltage
  ➢ Transient voltage

Copyright © 2008 by the Independent Electrical Contractors, Inc. All rights reserved.

Using Digital Multimeters to Diagnose Power Quality—Lab (Student Manual) Lesson 419-2

- Describe the general procedures for measurement of problems related to voltage change
  - Voltage sags and swells
  - Transient Voltage
- Discuss how harmonic distortion occurs, how it is classified, and how it is measured
- Discuss noise in a power distribution system and its effects
- Describe the general procedures for measuring, documenting and troubleshooting power quality problems
- Describe how phase unbalance can be a power quality problem
- Describe the significance of phase shift, inductance, capacitance, and impedance for power quality in AC circuits
- Discuss the effects of poor power quality
  - Heat
  - Skin Effect
  - Eddy Current
  - Circuit Breaker Tripping
  - Overheated Neutral Conductors
  - Overheated Conduit
  - Motor Problems
  - Transformer Problems
  - Lamp Problems
  - Electronic Equipment Problems
- Discuss general identification and troubleshooting procedures for power quality problems
- Describe how uninterruptible power systems work
  - Line Voltage Regulators
  - Surge Suppressors
  - Harmonic Filters
  - Line Filters
  - Power Conditioners
  - K-Rated Transformers
  - Isolation Transformers
  - Zig-Zag Transformers
- Discuss how the principles of proper electrical wiring and grounding insure good power quality and protect against power quality problems

Professionalism

- Work as a life-long learner to keep abreast of constant changes in the industry
- Ability to plan and organize learning materials
- Ability to manage tasks, resources and costs effectively
- Ability to plan and organize work projects

Using Digital Multimeters to Diagnose Power Quality—Lab (Student Manual) Lesson 419-3

Project Management

In the previous semester, you learned some basics about project management. The professionalism lessons in this semester will introduce you to some advanced aspects of project management. This is intended to continue to give you an overview of the important components of project management and how to use them. In addition to these mini-lessons, you should pay attention to how project management is done at your company. For instance, how does your company chart the progress of a project? How do you do a budget? How do you finish off a project? Paying attention to these things now will help you when you are asked to step up to the plate and become a project supervisor.

The professionalism lessons in this semester include:

- Planning for Time
- Scheduling Charts
- Planning Costs
- Controlling Work in Progress
- Project Control Charts
- Risk Management
- Monitoring Performance
- Taking Corrective Action
- Crisis Management
- Project Completion

Relevant Tools and Equipment

- *National Electrical Code®* 2008 edition
- Calculator
- Digital Multimeter
- Pictures and other examples of various electrical and telecommunications installations to illustrate the relationship between signal and power quality

Code References

- Article 100 Definitions
- Section 110 9 Interrupting Rating
- Section 110.10 Circuit Impedance and Other Characteristics
- Article 240 Overcurrent Protection
- Article 250 Grounding

Homework

📖 1) Reading Assignment

- None

AA 2) Key Terms

- None

📑 3) Practice Exercises

- None

Study Tips

▶ You will learn a lot of new concepts in this lesson. There are memory tips that can help you to remember all these terms.

- If you want to remember something, review it immediately, and then again within 20 minutes.

- Use a spaced practice schedule—20 minutes a day for five days is much better than two hours of memorization at one sitting.

- Make the information meaningful. Don't try to memorize information that you don't understand.

- Organize the information into chunks of seven items or less. If you have to memorize a big list of something, group into smaller lists organized in some logical fashion—even if only you know that logic.

▶ You will practice using digital multimeters to monitor power quality. Before you do the worksheets for this lab, be sure that you review the safety rules for using digital multimeters. If you have questions, be sure to ask your instructor before proceeding. Digital multimeters can be easily damaged and are expensive to replace.

Using Digital Multimeters to Diagnose Power Quality—Lab (Student Manual) Lesson 419-5

Tool Box Talk

Safety

✘ Before using any electrical test equipment, always refer to the user's manual for proper operating procedures, safety precautions and limits.

✘ After a demonstration, if a tester is passed around, operate only those parts that you have received permission to operate. Many of these testers are expensive and must be operated carefully to maintain their peak performance.

✘ Ensure that the test leads are connected properly. Test leads that are not connected to the correct jacks can be dangerous.

✘ Check test leads for frayed or broken insulation.

✘ Avoid taking measurements in humid or damp environments.

✘ Ensure that there are no atmospheric hazards such as flammable dust or vapor in the area.

Tricks of the Trade

✓ With any kind of testing, the testing equipment must have two characteristics: First, is it reliable? Second, are the results of the test valid? A reliable tester will show its reliability by repeatability or getting the same results with the same cable each time that it tests it. It will also get the same results on a calibration test using the same cable. You should perform a calibration test each time that you use a tester. Then you should perform the test again to make sure that you get the same results.

✓ You can check that the results of a tester are valid in several ways. This first is what is called a "sanity check." You can construct a sample length of cable and test it to check if the tester is providing the results that you think that it should. On more sophisticated testers, you can also run an auto test and ask the equipment to check itself. But an auto test should never replace the good thinking of the installer. Do you know the parameters of the cable and the limitations of the tester? Is the problem with the cable beyond the testing capabilities of the tester?

✓ Any piece of test equipment should be professionally calibrated on an annual basis.

Plans, Specifications, and Documentation

Be aware that testing equipment is also subject to recalls and product warnings. For example, OSHA, in August 2002, recalled about 650 electrical testing meters. An incompatible grommet, located in the battery compartment to protect internal wires, can cause the meter to provide inaccurate voltage and current readings. A meter that inaccurately indicates zero voltage or current creates the potential for electric shock or an electrocution hazard from the equipment being tested.

Review of National Electrical Code and Related Standards, Safety Requirements, and
Electrical Power Systems (Student Manual) Lesson 420-1

I E C NATIONAL	**Electrical Curriculum** **Year Four** **Student Manual**

Lesson 420 – Review of National Electrical Code and Related Standards, Safety Requirements, and Electrical Power Systems

Purpose

To review the *National Electrical Code® 2008* and related requirements pertaining to personal and workplace clearances, typical power systems, and low voltage systems in preparation for the journeyman's exam.

Objectives

By the end of this lesson, you should be able to:

420-1 Summarize the various codes and standards that will impact you as you pursue your trade as an electrician
420-2 Summarize the various types of injuries that can result from electrical shock and discuss the treatment for each
420-3 Identify and use appropriate electrical protective clothing
420-4 Describe the various types of voltage systems typically used in residential, commercial, and industrial applications
420-5 Interpret *Code®* requirements related to low voltage systems
420-6 Interpret *Code®* requirements related to working clearances

Copyright © 2008 by the Independent Electrical Contractors, Inc. All rights reserved.

Review of National Electrical Code and Related Standards, Safety Requirements, and Electrical Power Systems (Student Manual) Lesson 420-2

Content

Knowledge and Skills

- Electrical Codes, Standards, and Approvals—an Overview
  - NFPA 70- *National Electrical Code® 2008 (NEC®)*
    - History
    - Purpose
    - Applying the *NEC®*
    - Layout of the *NEC®*
  - OSHA Requirements
    - OSHA 1910, Subpart S-Electrical
    - OSHA 1926, Subpart K-Electrical
    - OSHA 1926, Subpart V-Power Transmission and Distribution
    - OSHA 1910, Subpart R-Special Industries
  - NFPA 70E-Electrical Safety Requirement for Employee Workplaces
  - NFPA 79-Electrical Standard for Industrial Machinery
  - NESC-National Electrical Safety Code
  - Third party certification
    - purpose of third part certification
    - UL White book
    - approval
  - Examination, installation, and use of equipment
    - examination
    - installation and use
  - Purpose of custom-made electrical equipment
  - Installing and maintaining electrical equipment
  - OSHA's relationship to electrical equipment
  - Purpose/Application of UL 508-Industrial Control Equipment
  - Authority Having Jurisdiction (AHJ)

- Electrical Safety and First Aid Review
  - The flow of electricity
  - Danger of electric shocks
  - Emergency response and first aid

- Dress Standards-Review
  - Head protection
  - Ear protection
  - Eyes and face protection
  - Hand and arm protection
  - Foot and leg protection
  - Overall protection

Review of National Electrical Code and Related Standards, Safety Requirements, and Electrical Power Systems (Student Manual) Lesson 420-3

- Electrical Systems
  - Systems of 600 volts or less
    - single phase systems
      - ☐ 120 volt, 2 wire
      - ☐ 208/120 volt, 3 wire
      - ☐ 240/120 volt, 3 wire
    - three phase systems
      - ☐ 208Y/120 volt, 4 wire
      - ☐ 240/120 volt, 4 wire
      - ☐ closed delta systems
      - ☐ open delta systems
      - ☐ 480Y/277 volt, 4 wire
  - Systems of 600 volts or more
    - 2.4kV to 15kV
    - 4,160 volts
    - 13,800 volts
  - Systems of less than 50 volts
    - systems installed in hazardous location
    - class 1, 2, & 3 circuits
    - fire-protective signaling circuits
    - nonpower-limited signaling circuits
    - power-limited conductor location
    - intrinsically safe systems
    - nonincendive circuits
    - luminary systems operating a 30 volts or less
- Working Clearances
  - Working space (600 volts or less)
    - working clearances
      - ☐ condition 1
      - ☐ condition 2
      - ☐ condition 3
    - clear spaces
    - access and entrance to working space
    - illumination
    - headroom
    - guarding of live parts

Review of National Electrical Code and Related Standards, Safety Requirements, and Electrical Power Systems (Student Manual) Lesson 420-4

- ➢ Working space (over 600 volts)
  - entrance and access to work space
  - work space
  - enclosure for electrical installations
  - installations accessible to qualified persons only
  - installations accessible to unqualified persons
  - workspace about equipment
    - ☐ front working space
      - ■ condition 1
      - ■ condition 2
      - ■ condition 3
  - separation from low-voltage equipment
  - locked rooms or enclosures
  - illumination
  - elevation of ungrounded live parts
- ➢ Identification of Disconnecting Means

Professionalism

- Ability to manage tasks, resources and costs effectively
- Ability to plan and organize work projects
- Ability to plan for time schedules

Review of National Electrical Code and Related Standards, Safety Requirements, and Electrical Power Systems (Student Manual) Lesson 420-5

Planning for Time

The idea when planning the amount of time for each project task is to determine the shortest time necessary to complete each task or step. Begin with the work breakdown structure and determine the time necessary to complete each step. Next, determine in what sequence the steps must be completed, and what steps can be under way at the same time. From this analysis, you will determine the three most significant time elements: (1) The duration of each step, (2) The earliest time at which each step may be started, (3) the latest time by which a step must be started.

Planning the time for each task needs to be done by people who have experience with the activities designated in each task. The experienced project manager can look at a task, understand what is required and provide a reasonable realistic "guesstimate" of the time required. You will get to that point someday if you start paying attention to the time it takes to accomplish things now. By paying attention, you will build up an understanding of the time involved. However, when it is your turn to build a schedule of work, if you are unsure of how long it takes to do something, you will need to rely on someone who does have the required experience.

Most project managers find it realistic to estimate time intervals as a range rather than as a precise amount. Because this is like trying to predict the future, you can only guess. But there are better ways to guess than others. When estimating the duration of tasks, there are six options that you can use to make the estimates as good as possible:

- Ask the people who will actually do the work, but have them estimate the work they'll be doing—not the durations. You will need to add extra time based on their workload, their other commitments and the experience with similar tasks.
- Get an objective expert's opinion (for example, someone who has worked on a similar project).
- Find a similar task in a completed project to see how long it took to get it done.
- If you have time and the task lends itself to this, perform a test session of the task to see how long it really takes using your resources.
- Make your best educated guess. This is the last resort when you are under pressure, so make sure the guess is based on as much experience as possible.
- Use an estimating resource such as R.S. Means.

Some project managers may want to start with the budget first, but that is usually not a good idea. Because time is money, the schedule will affect the final budget in many ways. That's why it is a good idea to talk about schedules first. It is best to try to determine how much time is really required to complete a project before you let your money (or lack of it) get in the way of your thinking.

Take some time to list some of the tasks from the project you are currently working on (or have just completed) and write down estimates of the time that you think it will take to complete each task.

Copyright © 2008 by the Independent Electrical Contractors, Inc. All rights reserved.

Relevant Tools and Equipment

- *National Electrical Code®*

Code References

- Article 411
- Article 502
- Article 504
- Article 725
- Article 110
- Article 501
- Article 503
- Article 720
- Article 760
- Chapter 9, Tables 11 A and 11 B

Homework

1) Reading Assignment

- Read *Stallcup's Electrical Design Book*, Chapters 1–5.

2) Key Terms

Using either the *Illustrated Dictionary for Electrical Workers*, the Glossary in the Appendix or your textbooks, write definitions for each of these terms before doing your homework:

- Alarm verification feature
- Auxiliary trip relay
- Channel
- Circuit interface
- Labeled
- Listed
- Master box
- Zone
- Annunciator
- Break glass station
- Chimes
- Communication channel
- Ground fault detector
- Trouble Signal

3) Practice Exercises

Answer the following referenced questions in the texts or other materials:

- In *Stallcup's Electrical Design Book* answer the odd numbered questions on pages 1-21, 1-22, 2-5, 3-5, 4-27, 4-28, 4-29, 4-30, 5-19, and 5-20.

Review of National Electrical Code and Related Standards, Safety Requirements, and Electrical Power Systems (Student Manual) Lesson 420-7

Study Tips

▶ Pay attention to the "Design Tips" found in *Stallcup's Electrical Design Book*. The tips provide important tips on the history and/or practical application of the *Code®* requirement under discussion.

Tool Box Talk

Safety

✗ Always assume that hazardous voltages exist in any wiring system. A safety check, using a known and reliable voltage measuring or detection device, should be made immediately before work begins and whenever work resumes. There are other important safety rules for avoiding electrical shock, some of which are given below.

1. Always use insulated tools and avoid all contact with bare terminals and grounded surfaces.

2. When cutting or drilling, be careful not to cut through or drill into concealed wiring or pipes. Make a small inspection opening before you start cutting.

3. When running wire on or near metallic siding, check for stray voltages. On mobile homes or trailers with metallic surfaces, always test for stray voltages and bond to ground before beginning work.

4. Be extra careful when working around other bare power wires or lightning rods, antennas, transformers, steam or hot-water pipes, and heating ducts.

Tricks of the Trade

✓ When planning for an electrical installation, project planners must anticipate where electrical circuits will run in crawlspaces, basements, and attics. They also need to know the direction of run for floor and ceiling joists. The electrical layout showing the distribution panel, convenience outlets, switches, and lights is usually placed on a copy of the floor plan and labeled "Electrical Plan." Broken lines indicate which outlets and switches are connected. However, the path of the wire is not necessarily where the lines are drawn.

✓ For larger construction jobs, an electrical plan may be prepared for outlets, another for lighting, and still another for the service entrance. These plans, together with the set of specifications, detail the electrical work to be done and the materials and fixtures to be used. For example, the Bay Colony Elementary School plans have individual plans for lighting and electrical power.

Copyright © 2008 by the Independent Electrical Contractors, Inc. All rights reserved.

Review of National Electrical Code and Related Standards, Safety Requirements, and Electrical Power Systems (Student Manual) Lesson 420-8

Plans, Specifications, and Documentation

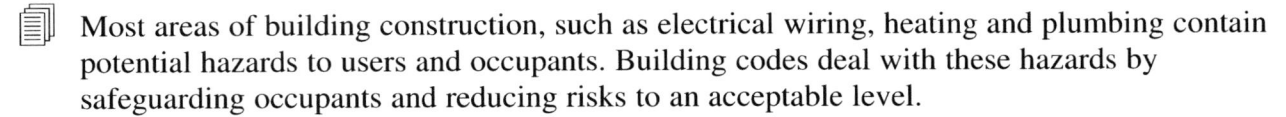

- Most areas of building construction, such as electrical wiring, heating and plumbing contain potential hazards to users and occupants. Building codes deal with these hazards by safeguarding occupants and reducing risks to an acceptable level.

- When beginning a project, be sure to check with the local municipal or county building department or inspector to find out what building code applies and if there are any exceptions. Building permits, fees and licensing could be discussed at the same time. You should also check with the state to determine if there are any state permits or licenses that apply to your installation, especially the state electrical board. It may also be useful to check with the general contractor if you are subcontracting outside your normal geographical area, and to talk to the potential customer if you are responding to a request for proposal.

Power Distribution (Student Manual) Lesson 421-1

| *I E C* NATIONAL | Electrical Curriculum

Year Four
Student Manual |

Lesson 421 – Power Distribution

Purpose

To familiarize the students with the systems used for power distribution, from electrical generation to electrical utilization.

Objectives

By the end of this lesson, you should be able to:

421-1 Identify the ways in which electricity is produced and distributed
421-2 Connect wye and delta transformer bank connections
421-3 Describe the functions of a substation
421-4 Explain the difference between feeders, busways, and other downstream systems
421-5 Discuss motor control centers and systems

Content

Knowledge and Skills

- Distribution System Overview
- Transformer Connections
  - Single-phase
  - Wye-wye
  - Wye-delta
  - Delta-delta
  - Delta-wye
  - Polarity observation

- Substations
  - Step up for transmission
  - Step down for utilization
  - Power disconnection
  - Voltage regulation
  - Voltage measurement
  - Switching point

- Switchboards and panelboards
  - Feeder distribution
  - Subfeeder distribution
  - Branch circuit distribution

- Motor control centers
  - Consolidation
  - Power
  - Control
  - Power transformation
  - Overload protection
  - Overcurrent protection

- Busways and other downstream systems
  - Power distribution
  - Convenience
  - Current capacity
  - Prefab fittings

Professionalism

- Ability to manage tasks, resources and costs effectively
- Ability to plan and organize work projects
- Ability to plan for time schedules
- Ability to use charts to schedule tasks

Scheduling Charts

There are several schedule-charting formats. The idea is to choose the format that best suits your project. Here are some options for displaying the time dimension of your project.

- Listings. These schedules simply list the tasks and the progress made on each task. These may not work well with very large, complex tasks that work interactively.
- Calendar charts. These are calendars with a space to make notes to keep track of task for small projects. Calendars are a good communication tool when displayed in a central location where all the team members can see the dates.
- Gantt charts. A Gantt chart is a horizontal bar graph that displays the time relationship of tasks in a project. A line placed on a chart in the time period during which the step should be done represents each step of a project. A Gantt chart shows a sequential flow of activities and activities that can be underway at the same time. Gantt charts are limited in their ability to show interdependence among tasks.
- PERT diagrams. PERT stands for Program Evaluation and Review Technique. It is a more sophisticated form of planning than a Gantt chart. It is appropriate for projects with many interactive tasks. A PERT diagram has three components: (1) Events represented by circles, (2) Activities represented by arrows connecting the events, (3) Non-activities connecting two events that are shown as dotted-line arrows. (A non-activity is a dependency between two events for which no work is required.) PERT diagrams are most useful if there is a specific amount of time scheduled for completing an activity.

Relevant Tools and Equipment

- None

Code References

- Article 100
- Article 384

Homework

1) Reading Assignment

- Read Chapter 11 in *Electrical Motors Controls for Integrated Systems* and do the questions on p. 270.

2) Key Terms

Using either the *Illustrated Dictionary for Electrical Workers*, the Glossary in the Appendix or your textbooks, write definitions for each of these terms before doing your homework:

- Alternator
- Busway
- Delta connection
- Motor control center
- NEMA configuration
- Panelboard
- Step up
- Step down
- Substation
- Switchboard
- Wye connection

3) Practice Exercises

Answer the following referenced questions in the texts or other materials:

- Complete Tech-Chek 11 and the Worksheets for Chapter 11 in *Workbook for Electrical Motor Controls for Integrated Systems*.

Study Tips

▶ This is one of the longer chapters in this book. Some of the information is review, but there is still a great deal of material for you to remember. It will be easier to study this chapter if you break it into several parts that you study at different times during the week.

Power Distribution (Student Manual) Lesson 421-5

Tool Box Talk

Safety

✗ The safety of all those who use a power distribution system depends upon the critical connection between grounding electrodes and grounding electrode conductors. Be sure that you understand the relationship between these two.

✗ The more grounding electrodes there are in a system, or the more area that one grounding electrode exposes to the earth, the better the electrode ground will be.

✗ In the case of multiple electrodes, one conductor can be used to connect all the electrodes in series, or separate grounding electrode conductors can be run to each grounding electrode.

✗ Grounding electrodes can be bonded together independent of the grounding electrode conductor as long as one of the bonded electrodes is attached to the grounding electrode system and the conductor is sized according to *NEC®* Section 250.66.

Tricks of the Trade

✓ Impedance is a key concept in understanding why ground faults can be so damaging. The fault allows a very low impedance path to ground. The equipment ahead of the short or the fault is the only limiting factor on the current. The circuit will attempt to draw current from other parts of the distribution system to meet the power demands of the fault. The current flow may instantly increase to over 100,000 amperes.

✓ With such a sharp rise in current there is also a sharp rise in the temperature of the conductor involved with the fault. The air around the fault will also become heated, ionized, and become a conductor rather than an insulator. Without the proper circuit design and protection, the electrical equipment and metal enclosures can melt and even explode. That is why every circuit must have circuit interruption protection that will not allow problems from a fault to exist more than a fraction of a second. Circuit breakers and fuses are rated to interrupt an electrical circuit if the current exceeds acceptable levels.

Plans, Specifications, and Documentation

Local building inspectors are typically electricians who have come up through the ranks. They know building construction and the *National Electrical Code®*. Don't hesitate to approach them directly and establish a positive relationship. Never just start to work and let the building inspector come on-site and catch you.

Services, Switchboards and Panelboards (Student Manual) Lesson 422-1

I E C
PRIDE
NATIONAL

Electrical Curriculum

Year Four
Student Manual

Lesson 422 – Services, Switchboards and Panelboards

Purpose

To review the *National Electrical Code®* pertaining to services, switchboards, and panelboards in preparation for the journeyman's exam.

Objectives

By the end of this lesson, you should be able to:

422-1   Interpret *Code®* requirements related to services
422-2   Interpret *Code®* requirements related to switchboards and panelboards

Content

Knowledge and Skills

- Services
  - Number of services permitted to a building (Article 230, part I).
    - Number of service drops or laterals
      - ☐ fire pumps
      - ☐ emergency luminary or power systems
      - ☐ multiple-occupancy building
      - ☐ large capacity requirements
      - ☐ building covering a large area
      - ☐ different characteristics
      - ☐ underground sets of conductors
    - Service must not pass through one building to supply another

- Overhead services (Article 230, Part II)
  - Insulation or covering
  - Size and rating
  - Clearances
    - ☐ Above roofs
    - ☐ Vertical clearance from ground
    - ☐ Point of attachment
  - Means of attachment
  - Service masts
  - Supports

- Underground Services (Article 230, Part III)
  - Insulation
  - Lateral size and rating

- Service-entrance conductors (Article 230, Part IV)
  - Number
    - ☐ Insulation
    - ☐ Size and rating
  - Individual conductors entering building or other structure
  - Raceways to drain
  - Overhead service locations
    - ☐ Service equipment termination
    - ☐ Identifying higher voltages to ground

- Service equipment general (Article 230, Part V)
  - Enclosed or guarded
  - Grounding and bonding

- Disconnecting means for service equipment (Article 230, Part VI).
  - Location
  - Marking
  - Maximum number
  - Grouping
  - Open all poles simultaneously
  - Grounded conductor
  - Rating of disconnect
    - ☐ One circuit installation
    - ☐ Two-circuit installation
    - ☐ One-family dwelling
  - Combined rating of disconnects
  - Equipment connected to the supply side of service disconnect

Services, Switchboards and Panelboards (Student Manual) Lesson 422-3

- ➢ Overcurrent protection for service equipment (Article 230, Part VII).
  - Where required
  - Location of
  - Specific circuits
  - Ground-fault protection
- ➢ Services exceeding 600 volts (Article 230, Part VII).

- Switchboards and Panelboards
  - ➢ Switchboard and panelboards (Article 408, Part I).
    - Support and arrangement of busbars and conductors
      - ☐ Conductors and busbars on a switchboard or panelboard
      - ☐ Overheating and inductive effects
      - ☐ Used as service equipment
      - ☐ High-leg marking
      - ☐ Phase arrangement
      - ☐ Minimum wire bending space
      - ☐ Circuit director

  - ➢ Switchboards (Article 408, Part II).
    - Location of switchboards
    - Clearances
    - Clearances for conductors entering bus enclosures
    - Grounding switchboard frames

  - ➢ Panelboards (Article 408, Part III).
    - Classification of panelboards
    - Number of overcurrent devices
    - Overcurrent protection
    - Damp and wet locations
    - Enclosure
    - Grounding of panelboards

Professionalism

- Ability to manage tasks, resources and costs effectively
- Ability to plan and organize work projects
- Ability to plan for time schedules
- Ability to use charts to schedule tasks
- Knowledge of budgets and budgeting

Planning Costs

A good plan involves doing research on sources of supplies and materials to assure that estimated costs are realistic. If you overestimate costs you may lose the job before you begin because your rates are not competitive. Some inaccuracies in budget are inevitable, but they should not be the consequence of insufficient attention while drafting your plan. The goal is to be as realistic as possible.

You cannot estimate the cost of your project until you know how long it will take, since labor is typically the highest cost item. Therefore, you should use your tasks and project schedule as the starting point for developing your project budget. Typical budget areas include:

- Labor. The wages paid to all staff working on the project for the time spent on the project.
- Equipment. The cost of purchasing or renting equipment for the project.
- Materials. The cost of items purchased for use in the project, such as cable, splicing kits, etc.
- Supplies. The cost of tools, office supplies and other general items needed by the project.
- Overhead. The cost of fringe benefits, usually a percentage of labor costs.
- General and Administrative. Cost of management and support services, e.g. payroll or purchasing.
- Profit. The reward to the firm for successfully completing the project, usually a percent of the cost.

With the cost components identified and the project broken down into steps, you can create a worksheet or spreadsheet to total the costs.

There are areas of potential budget problems that you need to be aware of. First, costs may go up between the time that you make up the budget and the time that you start the project. If you fail to get firm price commitments from suppliers and subcontractors, cost increases can cause problems for your budget. Secondly, poorly prepared project plans can lead to incomplete budgets. It is important to consider every task involved in getting the job done, every resource needed to accomplish that task and budget amounts for every one of these resources. Lastly, you need to make sure that every resource is available. For people, this means knowing their work availability, vacation schedule and commitment to other projects. For other resources, this means checking that all items will be available when you need them. If people or things are not available when you need them, it may cost you money that you didn't budget for.

Services, Switchboards and Panelboards (Student Manual) Lesson 422-5

Relevant Tools and Equipment

- *National Electrical Code® 2008*
- Calculator

Code References

- Article 100
- Article 230
- Article 408

Homework

1) Reading Assignment

- Read *Stallcup's Electrical Design Book*, Chapters 6 and 7.

2) Key Terms

Using either the *Illustrated Dictionary for Electrical Workers*, the Glossary in the Appendix or your textbooks, write definitions for each of these terms before doing your homework:

- High-leg
- Service drop
- Luminaries and appliance panelboard
- Service lateral
- Service point
- Power panelboard

3) Practice Exercises

Answer the following referenced questions in the texts or other materials:

- In *Stallcup's Electrical Design Book* answer the odd numbered questions on pages 6-35, 6-36, 6-37, 6-38, 6-39, 7-13, 7-14, and 7-15

Study Tips

▶ It is not too early to start studying for your Journeyman's Exam. Many of the lessons you have studied this semester cover the same topics as lessons you have had in previous years. In preparing for your Journeyman's Exam, start by reviewing all your notes, homework, glossaries of terms, and the exams and quizzes from each lesson. Is there anything that you are still uncertain about? If so, be sure to ask your instructor about these areas during the class session. Or, if you prefer, ask your instructor for more material to review on these topics.

Tool Box Talk

Safety

✗ Stay alert when working outside. Always locate all overhead lines. Assume that all overhead wires are energized and therefore dangerous. OSHA requires you to stay at least 10 feet away from overhead lines. This includes any object you could be handling. Following this widely adopted 10-foot rule will help keep you safe. Always be aware of HIGH VOLTAGE WARNING and DANGER signs and keep away.

✗ If you're moving ladders, long pipes or tall objects, trimming trees, painting with a long handle, installing antennas or moving sailboat masts, look up first to locate any overhead power lines and stay 10 feet away. Using tall objects near power lines can be extremely dangerous. Never lift them upright without first knowing what's above you. If you see an overhead line, stay more than 10 feet away. If you have a power line close to a tree that needs trimmed or pruned, contact a tree trimming professional.

✗ Occasionally, a severe storm can have the strength to blow down power lines. If you ever come across a downed power line, stay away. An energized line that has fallen across a car, a fence, a building or any other object can be very dangerous to unsuspecting persons. You should never touch a person, car, tree, limb or any other object that is in contact with a power line. Always assume a downed wire is dangerous and that it is energized. Touching a "live" line or anything near it—such as a fence, a puddle, a car, etc— can cause electricity to flow from it to you and through your body to reach the ground. This could result in serious injury and even death.

✗ If you come upon an accident that has caused a power line to fall onto a vehicle, stay away from the vehicle. Seek help (911) and tell the person(s) to wait in the car until the power company arrives. Remind them they are safe from electrical shock as long as they stay inside the car. If the occupant needs to escape because of danger (such as fire), the occupant must leap, landing with both feet together, jumping clear of the vehicle and never holding onto the door while leaping and once on the ground, must hop away, not run.

Tricks of the Trade

✓ Lightning strikes are rare, but fast rise time current comes from something that happens every day, every hour and every minute in every building. The cause of this fast rise time is called in-rush current. When you start up a motor, it draws six to 10 times its normal running current for one-half cycle to as long as 10 seconds. This in-rush current flowing down a conductor creates a voltage drop, which will lift local power circuits with respect to ground, or it can lift local ground with respect to other parts of a grounding structure for a building. This is not a problem unless you have a personal computer network with data cables strung around the building. If potential differences are not managed, they will present themselves across communications devices, data lines and computers in such a way that processing will be disrupted and equipment may be damaged. Remember that voltage drives current, and that electrical overstress resulting from pressure driving current through circuits can damage devices.

Plans, Specifications, and Documentation

- When specifying switchboards and panelboards on drawings, some engineers give only general specifications—how many volts, how many amps, amp interrupting capacity, whether or not it has a main breaker, copper or aluminum box, and how many of what size breakers go in it.

- Other times, the engineer will be more specific telling you exactly what brand and what breaker style you are to install. This may be because of a preference has developed over the years, or because of a specific customer requirement such as maintaining consistency with in a plant or throughout a chain of retail stores.

- If you want to install a panelboard of a different brand or style always communicate this through a submittal to the engineer and get his approval prior to procuring the equipment.

Conductors and Overcurrent Protection Devices (Student Manual) Lesson 423-1

IEC NATIONAL PRIDE

Electrical Curriculum

**Year Four
Student Manual**

Lesson 423 – Conductors and Overcurrent Protection Devices

Purpose

To review the *National Electrical Code®* requirements pertaining to the selection and installation of the proper type and size of conductors and overcurrent protection devices in preparation for the journeyman's exam.

Objectives

By the end of this lesson, you should be able to:

423-1   Identify conductor types and their intended application
423-2   Determine the ampacity of conductors
423-3   Compute proper circuit loading
423-4   Match temperature markings
423-5   Interpret and apply *NEC®* requirements for selecting and sizing conductors
423-6   Interpret and apply *NEC®* requirements for derating conductors
423-7   Interpret and apply *NEC®* requirements for protection of equipment conductors, and tap conductors
423-8   Select and apply circuit breakers and fuses in accordance with *NEC®* requirements
423-9   Interpret and apply *NEC®* requirements for ground fault protection for equipment and personnel

Content

Knowledge and Skills

- Conductors
  - Factors determining ampacity of conductors
    - Correction factors
    - Computing loads

Conductors and Overcurrent Protection Devices (Student Manual) Lesson 423-2

- Matching temperature marking
- Adjustment factors
- ➢ Definitions
- ➢ Current carrying conductors
  - Phase conductors
  - Neutral conductors
  - Control circuits
- ➢ Conductor identification
  - Grounded conductor
  - Equipment grounding conductor
  - Ungrounded conductors
- ➢ Conductors for general wiring
- ➢ Terminal ratings
- ➢ Long and short time ratings of conductors
- ➢ Derating circuit conductors
  - Over three current carrying conductors
  - Ambient temperature correction
- ➢ Diversity factors
- ➢ Individual circuits
- ➢ Conductors in parallel
- ➢ Wet locations
- ➢ Conductor constructions and applications
- ➢ Determining ampacities of conductors in electrical systems over 600 volts
  - Conductors in cables
  - Conductors in duct banks
  - Derating due to obstructions

- Overcurrent protection devices
  - ➢ Protection of equipment
  - ➢ Protection of conductors
    - Power loss hazard
    - Devices rated 800 amps or less
    - Devices rated over 800 amps
    - Tap conductors
    - Motor-operated appliance circuit conductors
    - Motor and motor control circuit conductors
    - Phase converter supply conductors
    - AC and refrigeration equipment circuit conductors
    - Transformer secondary conductors
    - Capacitor circuit conductors
    - Electric welder circuit conductors
    - Remote-control, signaling, and power limited circuit conductors
    - Fire alarm circuit conductors
  - ➢ Protection of fixture wires and cords
    - Flexible cord
    - Fixture wire
    - Listed extension cords

Conductors and Overcurrent Protection Devices (Student Manual) Lesson 423-3

- ➢ Standard Ampere Ratings
  - Fuses and fixed trip circuit breakers
  - Adjustable trip circuit breakers
- ➢ Supplementary overcurrent protection
- ➢ Electrical system coordination
- ➢ Using handle ties
- ➢ Protection of feeder and branch circuit taps (*NEC®* 240.21)
  - Making taps
  - Feeder taps 10 feet or less in length
  - Feeder taps up to 25 feet in length
  - Feeder taps including transformer
  - Feeder taps over 25 feet in length
  - Branch circuit taps
  - Busway taps
  - Motor circuit taps
  - Conductors from generator terminals
  - Tap from secondary of separately derived system
  - Outside feeder taps
  - Service conductors
- ➢ Grounded conductors
  - Common trip
  - Overload protection
- ➢ Change in size of grounded conductor
- ➢ Accessibility
  - Readily accessible
  - Accessible to occupant
- ➢ Fuses
  - Sizing
  - Time and temperature relationships
  - Markings
- ➢ Circuit Breakers
  - Sizing
  - Physical operation
  - Indicating
  - Markings
  - Applications
- ➢ Ground fault protection for equipment
  - Size
  - Setting
  - Clearing time
  - Written procedures
- ➢ GFCI for personnel
  - Function
  - Types
- ➢ OCPD for systems over 600 volts
  - Feeder circuit protection
  - Branch circuit protection

Copyright © 2008 by the Independent Electrical Contractors, Inc. All rights reserved.

Conductors and Overcurrent Protection Devices (Student Manual) Lesson 423-4

Professionalism

- Ability to manage tasks, resources and costs effectively
- Ability to plan and organize work projects
- Knowledge of budgets and budgeting
- Maintaining control of a project

Controlling Work in Progress

Once a project is underway, your key responsibility is to keep things going—on time and on budget. Getting a handle on your project isn't really hard. You have a plan and now is the time to use it. Your plan is the main tool for maintaining control of the project through its lifetime. This requires careful management of both the plan and its resources, including people. You will need to consistently monitor and update the plan. You should use the project specifications to monitor the quality of the project. You may need to update as conditions change so that it reflects the current status of the project, in terms of budget constraints, schedule changes or product modifications. You must monitor the progress of the project against the plan on a regular basis. Your boss will need to know about any significant departures from the budget or the schedule, otherwise these departures could affect the success of the project.

Remember that quality communication is a key to good project management. Every person involved in your project requires ongoing communication at various levels of detail. Not everyone needs to know everything, but everyone needs some information to help him or her do his or her part of the project. Some communication will be formal, some will be informal. The objective of communication is to keep people informed, on track, and involved in the project. It is up to you to get involved with all aspects of the project so that you keep track of what is going on in the project and contribute to the work being done.

Relevant Tools and Equipment

- *National Electrical Code® 2008*
- Calculator

Code References

- Article 100
- Article 240
- Article 310

Conductors and Overcurrent Protection Devices (Student Manual) Lesson 423-5

Homework

1) Reading Assignment

- Read *Stallcup's Electrical Design Book*, Chapters 8 and 9.

2) Key Terms

No new terms for this Lesson.

3) Practice Exercises

Answer the following referenced questions in the texts or other materials:

- In *Stallcup's Electrical Design Book* answer the odd numbered questions on pages 8-21, 8-22, 8-23, 9-33, 9-34, 9-35, 9-36, and 9-37.

Study Tips

▶ The best way to study the rest of the chapters in *Stallcup's Electrical Design Book* is to read each section in conjunction with the corresponding article in the *National Electrical Code®*. Read the *NEC®* first. Ask yourself if you could interpret and use the *NEC®* on the job. If you are not sure, make a list of questions that you need answered. Now read the corresponding section in *Stallcup's Electrical Design Book*. Did this answer your questions? If not, be sure to ask your instructor these questions during class.

Tool Box Talk

Safety

✗ Always check to see if conductors are damaged. Damaged conductors inside an electrical conduit might cause a short circuit. Both short circuits and ground faults are causes of very high current flow. If the neutral (grounded) conductor and one or more of the phase conductors have damaged insulation they might come into contact. The current would bypass the load but remain in the wiring system. Unlike short circuits, ground faults bypass the circuit conductor as a pathway back to the voltage source. Ground faults are particularly dangerous because the current leaves the normal electrical pathway.

Copyright © 2008 by the Independent Electrical Contractors, Inc. All rights reserved.

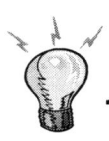

Tricks of the Trade

✓ In your electrical theory lessons, you learned that the higher the impedance, the lower the current flow at any given level of voltage. In other words, a low impedance path to ground will assure that fault current will be relatively large.

✓ In the event of either a short circuit or a ground fault, the current flow on the ground circuit will be large enough to exceed the current rating of the overcurrent protective device protecting that circuit. Once the current level is exceeded, the overcurrent device opens and shuts off the supply of current to the circuit, thus eliminating dangers from that fault. The low impedance path, in combination with the overcurrent device, protects the electrical system from excessive current levels due to faults.

✓ If an impedance path is too high, the current flow will be limited to levels that will not cause the overcurrent protection device to open. This will allow the fault to exist on the electrical system for a long period of time, a potentially fatal hazard. The aim of ground fault protection is to clear the fault as quickly as possible. If the current flow remains at a hazardous high level that causes no observable damage to equipment, burned-out fuses, or a circuit breaker trip, the circuit will remain energized. This will allow current to flow throughout the electrical system where it can come into contact with people. A low impedance ground path is critical to prevent this situation.

Plans, Specifications, and Documentation

Make sure you understand the difference between overloads, short circuits and ground faults. Overloads occur when there is a rise in current above the current rating of the wires in a circuit conductors. Short circuits occur when a high level of current flows between conductors. A ground fault exists when current flows from an ungrounded wire to ground. One overcurrent device usually protects at both levels but in the case of motor circuits, protection is often provided at two locations.

IEC Electrical Curriculum

Year Four
Student Manual

Lesson 424 – Grounding

Purpose

To review the *National Electrical Code®* requirement for grounding and bonding in preparation for the journeyman's exam.

Objectives

By the end of this lesson, you should be able to:

424-1   Interpret and apply *NEC®* requirements for circuit and system grounding
424-2   Interpret and apply *NEC®* requirements for equipment grounding
424-3   Interpret and apply *NEC®* requirements for supply and load equipment bonding
424-4   Interpret and apply *NEC®* requirements for grounding electrode systems

Content

Knowledge and Skills

- Equipment over 600 volts
  - Service point
  - Service conductors
  - Service entrance conductors
  - Warning signs
  - Isolating switches
  - Grounding connection
  - Disconnecting means
  - Circuit breakers
  - Arrestors
  - Service equipment
  - Metal-enclosed switch gear
  - Feeders supplying transformers
  - Supervised installations

- Services
- Feeders
- Branch circuits
- ➢ Grounding special equipment
- ➢ Wiring methods
  - Covers required
  - Conductors of different systems
  - Inserting conductors in raceways
  - Extra protection
  - Conductor bending radius
- ➢ Grounding
- ➢ Ground grid
- ➢ Designing feeder circuits

- Grounding
  - ➢ General requirements for grounding and bonding
  - ➢ Objectionable current over grounding conductors
  - ➢ Connection of grounding and bonding equipment
  - ➢ Protection of ground clamps and fittings
  - ➢ AC circuits and systems to be grounded
  - ➢ AC systems of 50 volts to 1000 volts not required to be grounded
  - ➢ Circuits not to be grounded
  - ➢ System grounding connections
  - ➢ Conductor to be grounded—AC systems
  - ➢ Main bonding jumper
  - ➢ Grounding separately derived AC systems
  - ➢ Two or more buildings or structures supplied from a common service
  - ➢ Portable and vehicle mounted generators
  - ➢ High impedance grounding
  - ➢ Grounding electrode system
  - ➢ Electrodes permitted for grounding
  - ➢ Supplementary grounding
  - ➢ Resistance of rod pipe and plate electrodes
  - ➢ Use of air terminals
  - ➢ Grounding electrode conductor material
  - ➢ Grounding electrode conductor installation
  - ➢ Size of alternating-current ground electrode conductor
  - ➢ Grounding electrode conductor and bonding jumper connection to grounding electrodes
  - ➢ Methods of grounding and bonding conductor connections to electrodes
  - ➢ Bonding
  - ➢ Services
  - ➢ Bonding for other systems
  - ➢ Bonding other enclosures
  - ➢ Bonding for over 250 volts
  - ➢ Equipment bonding jumpers
  - ➢ Bonding of piping systems and exposed structural steel
  - ➢ Types of equipment grounding, conductors
  - ➢ Identification of equipment grounding conductors

Grounding (Student Manual) Lesson 424-3

- Equipment grounding conductor installation
- Sizing of equipment grounding conductors
- Equipment grounding conductor connections
- Equipment fastened in place or connected by permanent wiring methods (fixed) grounding
- Cord and plug connected equipment
- Frames of ranges and clothes dryers
- Use of grounded circuit conductor for grounding equipment
- Connecting receptacle grounding terminal to box
- Continuity and attachment of equipment grounding conductors to boxes
- Swimming pools

Professionalism

- Ability to manage tasks, resources and costs effectively
- Ability to plan and organize work projects
- Ability to plan for time schedules
- Ability to use charts to schedule tasks
- Knowledge of budgets and budgeting
- Ability to use project control charts

Project Control Charts

A helpful tool for project management is a project control chart. This chart uses budget and schedule plans to give a quick status report of the project. It compares actual to planned, calculates a variance on task that is completed and tallies a cumulative variance for the whole project. To prepare a project control chart, list all the tasks for the project on a spreadsheet. Then use the schedule to list the time planned to complete each step and use the budget to list the expected cost of each task. As each task is completed, record its actual time and actual cost. Calculate the difference between the two. This is known as variance. Total all the variance to give you a quantity for how much the project is different from what was planned.

Another kind of project control chart is called a milestone chart. A milestone chart presents a broadbrush picture of a project's schedule and control dates. It lists those key events that require approval before the project can proceed or that need to be reported in communication with the bosses and/or the project team. If this is done correctly, a project will not have many milestones. Because of this lack of detail, a milestone chart is not very helpful during the planning phase when more information is needed. However, it is particularly useful in the implementation phase because it provides a concise summary of the progress that has been made. A milestone simply lists the task in the first column, the scheduled completion date in the second column and the actual completion date in the third column.

Budget control charts are generally of two varieties. One lists the project tasks with the actual costs compared to budgeted costs. It is similar to the project control charts, described above, and can be generated by hand or by computer. The other kind of budget control chart is a graph of budgeted costs compared to the actual costs. Either bar or line graphs may be used. Bar graphs usually relate budgeted and actual costs by project task, while line graphs usually relate planned cumulative project costs to actual costs over time. Another helpful approach to budget control is to compare the percentage of the project completed. The data can be compared by making a lilst or a graph. While the percentage of budget spent is a precise figure, the percentage of the project completed should be your best estimate of the project progress.

Relevant Tools and Equipment

- *National Electrical Code® 2008*
- Calculator

Grounding (Student Manual) Lesson 424-5

Code References

- Article 100
- Article 230
- Article 250
- Article 680
- NFPA 77
- NFPA 780

Homework

1) Reading Assignment

- Read *Stallcup's Electrical Design Book*, Chapters 10 and 11.

2) Key Terms

There are no new terms for this Lesson.

3) Practice Exercises

Answer the following referenced questions in the texts or other materials:

- In *Stallcup's Electrical Design Book* answer the odd numbered questions on pages 10-29, 11-77, 11-78, 11-79, 11-80, and 11-81.

Study Tips

▶ These chapters give you a number of problems with the steps to solve them and the answer to the problem. As you go through the chapters, try covering up the solution and seeing if you can arrive at the answer on your own. This will help you practice for the state electrical exam that you will take at the end of this apprenticeship program.

▶ Last week's lesson was on conductors and overcurrent protection devices. Review your homework and list the factors determining the ampacity of conductors and selection of the proper overcurrent protection device.

Copyright © 2008 by the Independent Electrical Contractors, Inc. All rights reserved.

Tool Box Talk

Safety

✘ Remember that three things determine the severity of electrical shock:

    1) Path of current through the body,
    2) Length of time that the current flows,
    3) The amount of current through the body.

✘ It takes less than one ampere of current to cause the heart to defibrillate. The best course of action when working around an electrical system is to make sure that power to the system is turned off completely.

Tricks of the Trade

✓ The following guidelines should be followed when installing a grounding conductor.

    1) Grounding electrode conductors should be as short as possible.
    2) Grounding electrode conductors should be as straight as possible.
    3) Avoid right angle bends in the conductor.
    4) No splices should exist in the conductor
    5) All connections should be free from corrosion.
    6) The grounding electrode conductors should be at least a #6 AWG conductor, as specified by the *NEC®*.
    7) All connections should be inspected once a year for tightness and completeness.

Plans, Specifications, and Documentation

NEC® Article 250 defines several unacceptable elements for grounding electrodes. These include the following.

    1) Aluminum electrodes are not acceptable because of the lower conductive properties of this metal.
    2) Metal underground gas piping systems are not acceptable because of the hazardous nature of these pipes.
    3) Interior water pipes located more than 1.5 meters (5 feet) from the entry point into a building are not acceptable because they are not easily accessible for inspection.

NEC® Article 250 also prohibits the use of insulated aluminum or copper clad aluminum as a grounding conductor where there is direct contact with the earth. These materials corrode easily when in contact with the earth.

Copyright © 2008 by the Independent Electrical Contractors, Inc. All rights reserved.

Why Buy Accubid?

estimating, billing, and project management software

- Accubid has over 20,000 users putting Accubid's estimating, project management, and service management software to the test in over 4,400 companies throughout North America
- Accubid estimating software is used by 9 out of the top 10 electrical contractors, and 36 out of the top 50*
- Accubid has been recognized as the estimating software of choice among specialty contractors in the last three CFMA surveys**
- Accubid has won the Technology's Hottest Companies award from Constructech magazine for the last four years running
- Accubid was the only software company to receive IEC's Innovative Product of the year award at the 2007 IEC National Conference
- Accubid is the only estimating software company to offer an unconditonal money-back guarantee
- Accubid is the only estimating software company to offer a 45-minute guaranteed support response time
- Accubid is the only software company where you can trade one product for another Accubid product

Discover for yourself why Accubid continues to be the contractor's choice.

* October 15, 2007 issue of Engineering News-Record (ENR) magazine.
** Statistics excerpted from the CFMA's 2002, 2004 and 2006 Information Technology Surveys for the Construction Industry with the permission of the Construction Financial Management Association, Princeton, NJ, 609-452-8000. CFMA neither evaluated nor ranked software in terms of performance. The survey should not be construed as the advice of CFMA.

Call for an online demonstration

total solutions for contractors

1-800-222-8243
www.accubid.com

ACCUBID

Designing and Installing Wiring Methods (Student Manual) Lesson 425-1

IEC NATIONAL

Electrical Curriculum

**Year Four
Student Manual**

Lesson 425 – Designing and Installing Wiring Methods

Purpose

To review the *National Electrical Code®* requirements pertaining to wiring methods in preparation for the journeyman's exam.

Objectives

By the end of this lesson, you should be able to:

425-1 Interpret and apply *NEC®* requirements for sizing boxes, conduit and fittings based on required conductor and device fill
425-2 Interpret and apply *NEC®* requirements for conductor fill for gutter, auxiliary gutters, panelboards, and cable tray
425-3 Interpret and apply *NEC®* requirements for box support
425-4 Interpret and apply *NEC®* requirements for AC, MC, NM, and SE cable
425-5 Interpret and apply *NEC®* requirements for RMC, IMC, and EMT
425-6 Interpret and apply *NEC®* requirements for flexible metal conduit and liquid tight flexible metal conduit
425-7 Interpret and apply *NEC®* requirements for the installation of cable and raceway systems

Content

Knowledge and Skills

- Box fill calculations
  - Conductor fill
  - Equipment grounding conductor fill
  - Device or equipment fill
  - Support fittings fill

Copyright © 2008 by the Independent Electrical Contractors, Inc. All rights reserved.

Designing and Installing Wiring Methods (Student Manual) Lesson 425-2

- Box volume calculations
  - Octagon boxes
    - Same size conductors
    - Different size conductors
  - Square boxes
    - Same size conductors
    - Different size conductors
  - Device boxes
    - Same size conductors
    - Different size conductors
  - Other boxes
    - Same size conductors
    - Different size conductors
  - Plaster rings and extension rings
- Conduit body sizing
- Junction box sizing
- Straight pull sizing
- Angle pull sizing
- Sizing conduits
  - Enclosing the same size conductors
  - Enclosing different size conductors
- Sizing nipples
- Gutter space sizing
- Auxiliary gutter sizing
- Panelboard gutter space requirements
- Cable tray sizing
  - Enclosing multi-conductor cables
  - Enclosing single conductor cables
  - Enclosing cable and signaling cables
- Boxes
  - Supporting boxes
  - Device boxes
  - Other boxes
  - Boxes used as supports
  - Without supporting devices or fixtures
  - Cable systems
  - Type AC armored cable
    - Uses permitted
    - Uses not permitted
    - Support

Designing and Installing Wiring Methods (Student Manual) Lesson 425-3

- ➢ Type MC metal-clad cables
  - Uses permitted
  - Uses not permitted
  - Support
- ➢ Nonmetallic sheathed cable (Romex)
  - Uses permitted
  - Uses not permitted
  - Installation
  - Through or parallel to framing members
  - Support
- ➢ Service-entrance cable
  - Uses permitted (service-entrance)
  - Uses permitted (branch circuits or feeder)
- ➢ Metal conduits
  - Intermediate metal conduit (IMC)
    - ☐ Uses permitted
    - ☐ Installation—size
    - ☐ Installation—reaming and threading
    - ☐ Couplings and connectors
    - ☐ Supports
    - ☐ Installation-bushings
  - Rigid metal conduit (RMC)
    - ☐ Installation—cinder fill
    - ☐ Installation—size
    - ☐ Installation—reaming and threading
    - ☐ Installation—bushings
    - ☐ Couplings and connections
    - ☐ Supports
  - Flexible metal conduit (Greenfield)
    - ☐ Uses permitted
    - ☐ Uses not permitted
    - ☐ Installation—grounding installation supports
  - Liquid tight flexible metal conduit (weatherproof flex)
    - ☐ Uses not permitted
    - ☐ Size
  - Rigid nonmetallic conduit (PVC)
    - ☐ Schedule 40
    - ☐ Schedule 80
    - ☐ Uses permitted
    - ☐ Uses not permitted
    - ☐ Installation—supports
  - Tubing
    - ☐ Uses permitted
    - ☐ Installation—size
    - ☐ Installation—couplings and connectors
    - ☐ Installation—supports

- Intermediate metal conduit (IMC)
  - ☐ Uses permitted
  - ☐ Installation—size
  - ☐ Installation—reaming and threading
  - ☐ Couplings and connectors
  - ☐ Supports
  - ☐ Installation—bushings
➢ Wiring method use in specific applications
  - Rough-in
  - Temporary pole
  - Slab
  - Walls
  - Cable systems in walls
    - ☐ Raceway systems in walls
  - Ceilings
    - ☐ Cable systems in ceilings
    - ☐ Raceway systems in ceilings
  - Attics
    - ☐ Cable systems in attics
    - ☐ Raceway systems in attics
  - Floors
➢ Flexible cords and extension cords
➢ Temporary wiring

Professionalism

- Ability to manage tasks, resources and costs effectively
- Ability to use charts to schedule tasks
- Knowledge of budgets and budgeting
- Ability to use project control charts
- Knowledge of risk management

Designing and Installing Wiring Methods (Student Manual) Lesson 425-5

Risk Management

You can reduce the risks on your project through risk management. Some people even refer to project management as the practice of risk management. In basic risk management, you plan for the possibility that a problem will occur by estimating the probability that the problem will arise during the project, evaluating the impact if the problem does arise, and preparing solutions in advance.

All risk management starts with identifying the risks. Identifying risks takes careful analysis. Assume that anything can go wrong. Learn from past projects. The failures of the past are often the best source of risk-control information. You should also examine critical relationships and what can happen if someone quits or doesn't deliver.

To analyze the probability that a risk will occur and the potential impact of the risk assign a number on a scale of 1 (lowest probability or impact) to 10 (highest probability or impact). Assign a number for both importance and for probability. Use this to help you determine the overall severity or importance of the risk, by multiplying the probability number by the impact number. This is a measure of severity. Usually an item with a number more than 40 requires careful analysis and monitoring.

Develop a response plan for the risks. You have four basic options for dealing with risks on your list:

1. Accept the risk.
2. Avoid the risk.
3. Monitor the risk and develop a contingency plan in case it appears that the risk will happen.
4. Transfer the risk, such as insurance or subcontracting.

Remember that risk management and response planning are ongoing processes. Risk must be regularly evaluated. Every project encounters problems neither planned for nor desired, and they will require some type of action before the project can continue.

Relevant Tools and Equipment

- *National Electrical Code® 2008*
- Calculator

Designing and Installing Wiring Methods (Student Manual) Lesson 425-6

Code References

- Article 314
- Article 320
- Article 330
- Article 334
- Article 338
- Article 342
- Article 344
- Article 348
- Article 350
- Article 352
- Article 358
- Article 392
- Article 400
- Chapter 9, Table 4
- Chapter 9, Table 5
- Annex C

Homework

1) Reading Assignment

- Read *Stallcup's Electrical Design Book*, Chapters 12 and 13.

2) Key Terms

There are no new terms for this Lesson.

3) Practice Exercises

Answer the following referenced questions in the texts or other materials:

- In *Stallcup's Electrical Design Book* answer the odd numbered questions on pages 12-31, 12-32, 12-33, 12-34, 13-25, and 13-26.

Study Tips

▶ There is an introduction at the start of each chapter in *Stallcup's Electrical Design Book*. Don't ignore these introductions. They have valuable information. Take a few minutes to review each of them.

▶ Last week's lesson was on lightning protection and grounding. Review your homework. Write in your own words, from memory, the definitions of bonding jumper, equipment grounding conductor, grounded conductor, and grounding electrode conductor. Check your definitions against those shown in Article 100 of the *NEC®*. If they are not correct, spend time reviewing Chapter 11 of *Stallcup's Electrical Design Book* and the definitions in Article 100 of the *NEC®*.

Copyright © 2008 by the Independent Electrical Contractors, Inc. All rights reserved.

Designing and Installing Wiring Methods (Student Manual) Lesson 425-7

Tool Box Talk

Safety

✘ Reading blueprints and drawings is an important part of working safely on the job. Drawings and specifications indicate the proper equipment and tools you will need for an installation. Architectural and structural drawings suggest the kinds of ladders you will need and ladder safety rules. Your working area will sometimes require special clothing or protective gear, and you can also "read" such special requirements from drawings and specifications. Of course, you should always wear proper clothing, including hardhat, safety glasses and gloves. Drawings can also tell you where there is risk of electrical exposure. Be sure that you follow electrical safety procedures when you work in such areas. It is a good idea to do a quick safety check to identify possible safety hazards using a set of drawings.

Tricks of the Trade

✓ When attempting to read blueprints on a job site make sure that you have a good location to place the plans. Plans are large and take up a lot of room. If you do not have adequate space to easily open and study the plans, or if there is inadequate lighting, details can easily be missed or misread. A special plan table is the best solution – get one if possible.

Plans, Specifications, and Documentation

If there is a discrepancy between what is on a set of construction drawings and what is written in the Specifications for a project, you should usually follow the Specifications information. Check with your employer or supervisor, but usually the specs have priority over the drawings.

AFCI Definitions from UL 1699 (Arc-Fault Circuit Interrupters) and History of *National Electrical Code® (NEC®)* AFCI Requirements

AFCI Definitions

Four of the relevant AFCI definitions from UL 1699 are:

Arc-Fault Circuit Interrupter (AFCI) - A device intended to mitigate the effects of arcing faults by functioning to de-energize the circuit when an arc-fault is detected.

Branch/Feeder Arc-Fault Circuit Interrupter - A device intended to be installed at the origin of a branch/circuit or feeder, such as at a panelboard. It is intended to provide protection of the branch/circuit wiring, feeder wiring, or both, against unwanted effects of arcing. This device also provides limited protection to branch circuit extension wiring. It may be a circuit-breaker type device or a device in its own enclosure mounted at or near a panelboard.

Outlet Circuit Arc-Fault Circuit-Interrupter – A device intended to be installed at a branch circuit outlet, such as at an outlet box. It is intended to provide protection of cord sets and power-supply cords connected to it (when provided with receptacle outlets) against the unwanted effects of arcing. This device may provide feed-through protection of the cord sets and power-supply cords connected to downstream receptacles.

Combination Arc-Fault Circuit Interrupter – An AFCI which complies with the requirements for both Branch/Feeder and outlet circuit AFCIs. It is intended to protect downstream branch circuit wiring and cord sets and power-supply cords.

National Electrical Code® (NEC®) Requirements

NEC 1999
The *National Electrical Code®* (*NEC®*) first mandated the use of AFCIs in 1999 for the protection of branch circuits supplying bedroom receptacle outlets. The effective date was January 1, 2002.

210-12. Arc-Fault Circuit-Interrupter Protection

    (A) Definition. An *arc-fault circuit interrupter* is a device intended to provide protection from the effects of arc faults by recognizing the characteristics unique to arcing and by functioning to de-energize the circuit when an arc fault is detected.
    (B) Dwelling Unit Bedrooms. All branch circuits that supply 125-volt, single phase, 15- and 20-ampere receptacle circuits installed in dwelling unit bedrooms shall be protected by an arc-fault circuit interrupter(s). This requirement shall become effective January 1, 2002.

NEC 2002
Protection was broadened from bedroom receptacles to bedroom outlets. There was additional stress on protection for the entire branch circuit.

210-12. Arc-Fault Circuit-Interrupter Protection

    (A) Definition. An *arc-fault circuit interrupter* is a device intended to provide protection from the effects of arc faults by recognizing the characteristics unique to arcing and by functioning to de-energize the circuit when an arc fault is detected.
    (B) Dwelling Unit Bedrooms. All branch circuits that supply 125-volt, single phase, 15- and 20-ampere outlets installed in dwelling unit bedrooms shall be protected by an arc-fault circuit interrupter listed to provide protection of the entire branch circuit.

NEC 2005

Combination AFCIs were mandated for the protection of branch circuits supplying bedroom outlets, with the continued use of Branch/Feeder AFCIs being permitted until January 1, 2008. An exception, with requirements, was added for AFCIs located within 1.8 m of the loadcenter.

210-12. Arc-Fault Circuit-Interrupter Protection

(A) **Definition**. An *arc-fault circuit interrupter* is a device intended to provide protection from the effects of arc faults by recognizing the characteristics unique to arcing and by functioning to de-energize the circuit when an arc fault is detected.

(B) **Dwelling Unit Bedrooms**. All 120-volt, single phase, 15- and 20-ampere branch circuits supplying outlets installed in dwelling unit bedrooms shall be protected by a listed arc-fault circuit interrupter, combination type installed to provide protection of the branch circuit.

      Branch/feeder AFCIs shall be permitted to be used to meet the requirements of 210.12(B) until January 1, 2008.

FPN: For information on types of Arc-Fault Circuit Interrupters, see UL 1699 – 1999, *Standard for Arc-Fault Circuit Interrupters.*

Exception: The location of the arc-fault circuit interrupter shall be permitted to be at other than the origination of the branch circuit in compliance with (a) and (b):

(a) *The arc-fault circuit interrupter installed within 1.8 m (6 ft) of the branch circuit overcurrent device as measured along the branch circuit conductors.*
(b) *The circuit conductors between the branch circuit overcurrent device and the arc-fault circuit interrupter shall be installed in a metal raceway or a cable with a metallic sheath.*

NEC 2008
The use of Combination AFCIs was expanded to many additional dwelling unit locations. Under specified conditions, Combination AFCIs were permitted to be located at the first outlet.

210-12. Arc-Fault Circuit-Interrupter Protection

(A) **Definition**. An *arc-fault circuit interrupter* is a device intended to provide protection from the effects of arc faults by recognizing the characteristics unique to arcing and by functioning to de-energize the circuit when an arc fault is detected.

(B) **Dwelling Units** All 120-volt, single-phase, 15- and 20- ampere branch circuits supplying outlets installed in dwelling unit family rooms, dining rooms, living rooms, parlors, libraries, dens, bedrooms, sun rooms, recreation rooms, closets, hallways, or similar rooms or areas shall be protected by a listed arc-fault circuit interrupter, combination-type, installed to provide protection of the branch circuit.

FPN: For information on types of Arc-Fault Circuit Interrupters, see UL 1699 – 1999, *Standard for Arc-Fault Circuit Interrupters.*

Exception: Where RMC, IMC or EMT or steel armored cable, Type AC, meeting the requirements of 250.118, using metal outlet or junction boxes is installed for the portion of the branch circuit between the branch circuit overcurrent device and the first outlet, it shall be permitted to install a combination AFCI at the first outlet to provide protection for the remaining portion of the branch circuit. The AFCI installed at the first outlet shall also provide protection for any equipment connected to that outlet.

Branch Circuits and Feeder Circuits (Student Manual) Lesson 426-1

IEC NATIONAL PRIDE

Electrical Curriculum

Year Four
Student Manual

Lesson 426 – Branch Circuits and Feeder Circuits

Purpose

To prepare students for the Journeyman's exam through review of the *National Electrical Code®* requirements pertaining to branch and feeder circuits.

Objectives

By the end of this lesson, you should be able to:

426-1   Interpret and apply *NEC®* requirements for residential branch circuits
426-2   Interpret and apply *NEC®* requirements for commercial and industrial branch circuits
426-3   Interpret and apply *NEC®* requirements for feeder circuits
426-4   Interpret and apply *NEC®* requirements for AC, MC, NM, and SE cable
426-5   Interpret and apply *NEC®* requirements for RMC, IMC, and EMT
426-6   Interpret and apply *NEC®* requirements for flexible metal conduit and liquid tight flexible metal conduit
426-7   Interpret and apply *NEC®* requirements for the installation of cable and raceway systems

Content

Knowledge and Skills

- Branch circuits
  - Residential
    - General purpose
    - Small appliance
    - Laundry
    - Individual
    - Residential cooking equipment
    - Dryer equipment
    - Rating

Branch Circuits and Feeder Circuits (Student Manual) Lesson 426-2

- Permissible loads
  - ☐ 15 and 20 amp branch-circuits
  - ☐ 30 ampere branch-circuits
  - ☐ 40 and 50 amp branch-circuits
  - ☐ branch-circuits larger than 50 amps
- Conductors
  - ☐ 60 degree Celsius rating
  - ☐ 75 degree Celsius rating
  - ☐ 90 degree Celsius rating
  - ☐ derating due to ambient temperature
- Voltage limitations
  - ☐ 120 volts between conductors
  - ☐ 277 volts-to-ground
  - ☐ 600 volts between conductors
- Amperage
  - ☐ Finding amperage for single phase circuits
  - ☐ Finding amperage for three-phase circuits

➢ Commercial and industrial
- Lightning loads
  - ☐ Noncontinuous operation
  - ☐ Continuous operation
- Other loads
- Electric discharge loads
- Outlets
- Show window loads
- Luminary track loads
- Sign loads
  - ☐ Size required
  - ☐ Rating
  - ☐ Computed load
- Receptacle loads
- General purpose
- Individual
- Multiwire branch-circuits
- Commercial cooking equipment
- Water heater loads
  - ☐ Conductors
  - ☐ Overcurrent protection
  - ☐ Disconnecting means
- Heating loads
  - ☐ Conductors
  - ☐ Overcurrent protection
  - ☐ Disconnecting means
- Air conditioning loads
  - ☐ Conductors
  - ☐ Overcurrent
  - ☐ Disconnecting means

Copyright © 2008 by the Independent Electrical Contractors, Inc. All rights reserved.

Branch Circuits and Feeder Circuits (Student Manual) Lesson 426-3

- Motor loads
  - ☐ Conductors
  - ☐ Overcurrent protection
  - ☐ Disconnecting means
- Welder loads
  - ☐ AC/DC are welders
  - ☐ Conductors
  - ☐ Overcurrent protection
- Motor-generator arc welders
  - ☐ Conductors
  - ☐ Overcurrent protection
- Resistance welders
  - ☐ Conductors
  - ☐ Overcurrent protection

- Feeder circuits
  - Loads
  - Voltage
  - Ungrounded phase conductors
  - Computing amps
  - Power factor
  - Actual power single-phase
  - Noncontinuous operated loads
  - Continuous operated loads
  - Demand factors
    - Applying demand factors for receptacle loads
  - Neutral
    - Sizing calculating neutral current for three-phase currents
    - Harmonic currents and effects on neutral
    - Sizing neutral elements
    - Demand factors
  - Voltage drop
    - Single-phase circuits
    - Three phase circuits
  - Sizing the equipment grounding conductor
  - Sizing the neutral conductor
  - Sizing conductors based on voltage drop
  - Calculation of a feeder circuit
  - Optional calculation for an existing feeder circuit

Professionalism

- Ability to manage tasks, resources and costs effectively
- Ability to plan and organize work projects
- Ability to plan for time schedules
- Ability to use charts to schedule tasks
- Knowledge of budgets and budgeting
- Ability to use project control charts
- Knowledge of risk management
- Willingness to monitor progress on a project

Monitoring Performance

The heart of project management is monitoring work in progress. It is your way to know "what is going on" and how "what is going on" compares to what was planned for. With effective monitoring, you will know if and when corrective action is required. Since the status is constantly changing, you'll need to monitor the project and compare to the plan in some way every day. Reports are good tools for combining information, but informal discussions often reveal a more accurate picture of the project.

Following are ways to keep abreast of project progress:

- Inspection is probably the most common way to monitor project progress. Inspection is an effective way to see whether project specifications are being met, as well as whether there are unnecessary waste or unsafe work practices. Ask questions and listen to explanations.

- Interim progress reviews are communications between the project manager and those responsible for various steps of a project. Progress reports typically occur on a fixed time schedule or at the completion of each task. Three topics are usually covered during these reviews: (1) Review of the progress compared to the plan, (2) Review of problems encountered and how they were handled, (3) review of anticipated problems with proposed plans for handling them.

- Testing is another way to verify project quality. Certain tests are usually written in to the specifications to confirm that the desired quality is achieved.

Auditing can be done during the course of the project or at its conclusion. Common areas for audit are financial, maintenance procedures, purchasing practices, and safety practices. Auditors should be expert in the area under review.

Branch Circuits and Feeder Circuits (Student Manual) Lesson 426-5

Relevant Tools and Equipment

- *National Electrical Code® 2008*
- Calculator

Code References

- Article 210
- Article 215
- Article 220

Homework

1) Reading Assignment

- Read *Stallcup's Electrical Design Book*, Chapters 14 and 15.

2) Key Terms

There are no new terms for this Lesson.

3) Practice Exercises

Answer the following referenced questions in the texts or other materials:

- In *Stallcup's Electrical Design Book* answer the odd numbered questions on pages 14-39, 14-40, 14-41, 14-42, 15-17, and 15-18.

Study Tips

▶ As you study for class, try to relate the issues you are studying to specific situations you have encountered on the job. This will help you remember what you have studied. Think about problems you might have had in those situations; and how you might have handled them differently. Write down any questions that arise so you can ask your instructor.

Tool Box Talk

Safety

✘ Electrical power must be turned off whenever electrical equipment is installed, serviced, inspected, or repaired. The power must be removed and the equipment must be locked out and tagged out. Lockout is the process of removing the source of electrical power and installing a lock that will prevent the power from being turned "on." Tag out refers to the placement of a danger tag on the electrical power sources to indicate that that the equipment should not be operated until the danger tag is removed.

✘ Lock out/tag out procedures are used whenever equipment is serviced and does not require power to be turned "on" to perform the service. It should also be used if machine guards or safety devices are bypassed or removed, or if any possibility of injury exists if the power is turned "on." Finally, lock out/tag out should be used whenever jammed equipment needs to be cleared.

✘ There are a variety of lock out devices designed to fit standard electrical controls. Different kinds of warning tags are also available with self-locking, non-reusable tag ties. Be sure to be aware of them and heed them on the worksite.

Tricks of the Trade

✓ Manufacturers may give load sizing for uninterruptible power supply (UPS) devices in either watt (W) or volt-ampere (VA) measurement units. The volt-ampere rating system is better for matching the load to the UPS because the fundamental factor that limits the output capacity of a UPS is its output current, which is more closely related to volt-amperes than to watts.

Plans, Specifications, and Documentation

▤ As you go through the blueprints to install wiring systems remember that One-Line and Riser diagrams contain a lot of information. Practice using One-Line Diagrams as often as possible and take the time to study them until you are familiar with all the information they contain.

▤ Remember that plans for the Bay Colony Elementary School do not show the routing for the conduits for the lighting and receptacle outlets, although the circuiting is shown on the plans. When routing conduit or cable for wiring branch circuits, the circuit each outlet is installed on is important, and should not be changed unless directed to do so by a supervisor.

Give us a journeyman

We'll give you an estimator

Can't find qualified electrical estimators? Well, stop looking outside and start looking within your own organization. Sign up a journeyman for the Accubid Certified Estimator (ACE) program, which blends professional development courses with Accubid software training labs to give students the technical skills they need to excel at computerized estimating. After successfully completing the 5-part program, he'll be officially designated as an Accubid Certified Estimator*.

What can you or your staff learn from Accubid's Computerized Estimating courses?
- Project initiation procedures
- Understanding specifications
- Understanding estimate setup procedures
- How to break down a project
- Systematic takeoff and verification
- Understanding the bid closing procedure to arrive at a job cost
- Understanding computerized estimating software
- Takeoff modification procedures
- Advanced sorting and filtering procedures
- How to analyze labor, material, equipment, and other costs
- How to validate an estimate to ensure accuracy
- Gain new insights into the estimating process

For a free brochure, or to register for a course, call 1-800-ACCUBID (222-8243) or visit our website at www.accubid.com.

*To be designated as an Accubid Certified Estimator, the student must successfully complete the following courses and training labs: Computerized Estimating Level 100 (Takeoff), Accubid Estimating Software Training (Beginner), Computerized Estimating Level 200 (Closing), Accubid Estimating Software Training (Advanced), and Computerized Estimating Level 300 (Examination).

ACCUBID education

Professional Development for Contractors
1-800-222-8243 • www.accubid.com

Mid-Term Review and Exam (Student Manual) Lesson 427-1

| **I E C**
PRIDE
NATIONAL | **Electrical Curriculum**

Year Four
Student Manual |

Lesson 427 – Mid-Term Review and Exam

Purpose

To prepare students for the Journeyman's exam through a comprehensive review of *National Electrical Code®* requirements pertaining to the selection and installation of the proper type and size of conductors and overcurrent protection devices.

Objectives

By the end of this lesson, you should be able to:

419-1   Describe major power quality problems and sources
419-2   Discuss the basic concepts and procedures for power quality measurement
419-3   Discuss the effects of poor power quality on electrical equipment and systems
419-4   Discuss indicators of power quality problems and the equipment and methods used to resolve those problems

420-1   Summarize the various codes and standards that will impact you as you pursue your trade as an electrician.
420-2   Summarize the various types of injuries that can result from electrical shock and discuss the treatment for each.
420-3   Identify and use appropriate electrical protective clothing.
420-4   Describe the various types of voltage systems typically used in residential, commercial, and industrial applications.
420-5   Interpret *Code®* requirements related to low voltage systems.
420-6   Interpret *Code®* requirements related to working clearances.

421-1   Identify the ways in which electricity is produced and distributed.
421-2   Connect wye and delta transformer bank connections.
421-3   Describe the functions of a substation.
421-4   Explain the difference between feeders, busways and other downstream systems.
421-5   Discuss motor control centers and systems.

Mid-Term Review and Exam (Student Manual) Lesson 427-2

422-1	Interpret *Code®* requirements related to services.
422-2	Interpret *Code®* requirements related to switchboards and panelboards.
423-1	Identify conductor types and their intended application.
423-2	Determine the ampacity of conductors.
423-3	Compute proper circuit loading.
423-4	Match temperature markings
423-5	Interpret and apply *NEC®* requirements for selecting and sizing conductors.
423-6	Interpret and apply *NEC®* requirements for derating conductors.
423-7	Interpret and apply *NEC®* requirements for protection of equipment conductors, and tap conductors.
423-8	Select and apply circuit breakers and fuses in accordance with *NEC®* requirements
423-9	Interpret and apply *NEC®* requirements for ground fault protection for equipment and personnel.
424-1	Interpret and apply *NEC®* requirements for circuit and system grounding.
424-2	Interpret and apply *NEC®* requirements for equipment grounding.
424-3	Interpret and apply *NEC®* requirements for supply and load equipment bonding.
424-4	Interpret and apply *NEC®* requirements for grounding electrode systems.
425-1	Interpret and apply *NEC®* requirements for sizing boxes, conduit and fittings based on required conductor and device fill.
425-2	Interpret and apply *NEC®* requirements for conductor fill for gutter, auxiliary gutters, panelboards, and cable tray.
425-3	Interpret and apply *NEC®* requirements for box support.
425-4	Interpret and apply *NEC®* requirements for AC, MC, NM, and SE cable
425-5	Interpret and apply *NEC®* requirements for RMC, IMC, and EMT.
425-6	Interpret and apply *NEC®* requirements for flexible metal conduit and liquid tight flexible metal conduit.
425-7	Interpret and apply *NEC®* requirements for the installation of cable and raceway systems.
426-1	Interpret and apply *NEC®* requirements for residential branch circuits.
426-2	Interpret and apply *NEC®* requirements for commercial and industrial branch circuits.
426-3	Interpret and apply *NEC®* requirements for feeder circuits.
426-4	Interpret and apply *NEC®* requirements for AC, MC, NM, and SE cable.
426-5	Interpret and apply *NEC®* requirements for RMC, IMC, and EMT.
426-6	Interpret and apply *NEC®* requirements for flexible metal conduit and liquid tight flexible metal conduit.
426-7	Interpret and apply *NEC®* requirements for the installation of cable and raceway systems.

Copyright © 2008 by the Independent Electrical Contractors, Inc. All rights reserved.

Mid-Term Review and Exam (Student Manual) Lesson 427-3

Content

Knowledge and Skills

- Power quality and troubleshooting
- Power distribution
- *NEC®* and related requirements related to personal and workplace safety, workplace clearances, typical power systems and low voltage systems.
- Services, switchboard, and panelboards
- Conductors and overcurrent protective devices
- Lightning protection and grounding
- Wiring methods
- Brand Circuits and feeder circuits

Professionalism

- Ability to manage tasks, resources and costs effectively
- Ability to plan and organize work projects
- Ability to plan for time schedules
- Ability to use charts to schedule tasks
- Knowledge of budgets and budgeting
- Ability to use project control charts
- Knowledge of risk management
- Willingness to monitor progress on a project
- Willingness to take corrective action to ensure the success of a project

Taking Corrective Action

As a project progresses and you monitor performance, there will be times when the project does not measure up to the plan. This calls for corrective action. However, don't be too quick to take action. Some deficiencies are self-correcting. It is unrealistic to expect steady and consistent progress day after day. Sometime you fall behind and sometime you'll be ahead. The successful project manager is able to maintain a perspective that allows honest, objective evaluation of each issue as it comes up. The key is to monitor things so that nothing gets out of hand.

When quality is not according to specification, the customary action is to do it over, according to plan. However, if the work or material exceeds specifications, you or your client may choose to accept it.

When the project begins to fall behind schedule, there are three alternatives that may correct the problem. The first is to exam the work remaining to be done and decide whether the lost time can be recovered in the next tasks. If this is not feasible, your company may want to offer an incentive for on-time completion of the project. The incentive could be justified if you compare the cost to potential losses due to late completion. Sometimes a supplier can deliver a partial order to keep your project on schedule and complete the delivery later. Finally, consider bringing more resources. This too will cost more, but may offset further losses from delayed completion.

When a project begins to exceed budget, consider the work remaining and whether or not cost overruns can be recouped on work yet to be completed. If this isn't practical, consider narrowing the project scope or renegotiate with the client. Perhaps nonessential elements of the project can be eliminated, thereby reducing coasts or saving time. When something is not available or is more expensive than budgeted, substituting a comparable item may solve your problem. Or you may need to look for other suppliers that can deliver within your budget and schedule. Sometimes demanding that people do what they agreed to do gets the desired results. You may have to appeal to higher management for backing and support.

Relevant Tools and Equipment

- *National Electrical Code® 2008*
- Calculator

Mid-Term Review and Exam (Student Manual) Lesson 427-5

👉 Code References

- Article 100
- Article 110
- Article 210
- Article 215
- Article 220
- Article 230
- Article 240
- Article 250
- Article 310
- Article 314
- Article 320
- Article 330
- Article 334
- Article 338
- Article 342
- Article 344
- Article 348
- Article 350
- Article 352
- Article 358
- Article 392
- Article 400
- Article 408
- Article 501
- Article 502
- Article 503
- Article 504
- Article 680
- Article 720
- Article 725
- Article 760
- Chapter 9, Table 4
- Chapter 9, Table 5
- Annex C

Homework

1) Reading Assignment

- Review all your homework assignments for Lessons 419-426.

2) Key Terms

Review all the terms from Lesson 419-426.

3) Practice Exercises

- None

Study Tips

▶ These are a few hints to remember when you take tests like the Mid-Term Exam:

A. Be sure that you understand the instructions.

B. Read all the questions first. Then answer the easiest questions first. This will help build your confidence and ensure that you get credit for the easy questions.

C. Don't struggle with a question and waste your time. Go on and come back to it later.

D. Make sure that you allow enough time to finish all the questions.

E. If you are unsure of an answer, put down what you think is the best answer and come back to it later. Something in the rest of the test may trigger the correct answer for you.

Receptacles, Luminaires, and Switching Outlets (Student Manual) Lesson 428-1

IEC NATIONAL PRIDE

Electrical Curriculum

Year Four
Student Manual

Lesson 428 – Receptacles, Luminaires, and Switching Outlets

Purpose

To prepare students for the Journeyman's exam through comprehensive review of the *National Electrical Code®* requirements pertaining to receptacles, luminaires, and switching outlets.

Objectives

By the end of this lesson, you should be able to:

428-1   Interpret and apply *NEC®* requirements for residential receptacles outlets
428-2   Interpret and apply *NEC®* requirements for GFCI protection of residential receptacle outlets
428-3   Interpret and apply *NEC®* requirements for GFCI protection of receptacles in and around boathouses, swimming pools, storage pools, spas, hot tubs, and hydromassage tubs
428-4   Interpret and apply *NEC®* requirements for receptacles on construction sites
428-5   Interpret and apply *NEC®* requirements for receptacles installed in commercial and industrial locations
428-6   Interpret and apply *NEC®* requirements for residential luminary and switching outlets
428-7   Interpret and apply *NEC®* requirements for commercial and industrial luminary and switching outlets
428-8   Interpret and apply *NEC®* requirements for switching outlets by pools, hot tubs, and spas

Content

Knowledge and Skills

- Receptacle Outlets
  - Grounding receptacles
  - Non-grounding receptacles
  - Replacement of receptacles

Copyright © 2008 by the Independent Electrical Contractors, Inc. All rights reserved.

- Dwelling unit locations
- Wall receptacle outlets
- Receptacle outlets for small appliance circuits
- Receptacle outlets over countertops
- Wall type countertops
- Peninsular countertops
- Island countertops
- Receptacle outlets in bathrooms
- Receptacle outlets outdoors
- Laundry circuits
- Receptacle outlets in basements and garages
- Receptacles in basements
- Receptacles in garages
- Receptacles in hallways
- GFCI protection of receptacles
    - In bathrooms
    - In garages
    - Outdoors
    - In basements and crawlspaces
    - Over countertops for boathouses
    - Around swimming pools
    - Storable pools
    - Spas or hot tubs
    - Hydromassage tubs
    - On construction sites
    - Bathrooms in commercial and industrial locations
    - Rooftops of commercial and industrial buildings
    - Used for maintenance activities
- Luminaire and Switching Outlets
    - Grounding luminaires (fixtures)
    - Ungrounded luminaires (fixtures)
    - Replacement luminaires (fixtures)
    - Number on a circuit
        - Residential
        - Commercial/industrial
    - Luminaire outlets in dwelling units
        - In habitable rooms
        - In hallways
        - In bathrooms
        - Wall-switched receptacle luminary outlets
        - In garages
        - In utility rooms
        - At outside doors
        - In basements
        - In attics
        - In underfloor spaces
        - Luminaire outlets over bathtubs

Receptacles, Luminaires, and Switching Outlets (Student Manual) Lesson 428-3

- Luminaire outlets in clothes closets
  - ☐ Storage space
  - ☐ Types permitted
  - ☐ Types not permitted
- ➢ Guest rooms in hotels and motels
- ➢ Commercial and industrial attics and crawl spaces
- ➢ Illumination for electrical equipment in commercial and industrial locations
- ➢ Recessed luminaires (fixtures) discharge luminary
- ➢ Discharge luminaires
- ➢ Luminaire outlets at swimming pools
  - Underwater luminaires (fixtures)
  - Wet-niche
  - Dry-niche
  - No-niche
- ➢ Luminaire outlets over spas or hot tubs
- ➢ Luminaire outlets over hydromassage tubs
- ➢ Switching outlets
  - Types of switches
  - Switching outlet heights
  - Loading switches
  - By swimming pools
  - By spas or hot tubs
  - Grounded or ungrounded

Professionalism

- Ability to manage tasks, resources and costs effectively
- Ability to plan and organize work projects and time schedules
- Ability to use charts to schedule tasks and use project control charts
- Knowledge of budgets, budgeting, and risk management
- Willingness to monitor progress on a project
- Willingness to take corrective action to ensure the success of a project
- Understanding the need for crisis management

Crisis Management

Conflicts are a way of life in projects. Sometimes project managers are described as conflict managers. But conflicts don't have to bring your project to a halt. If you know that something is going to happen, you can plan for dealing with it. Recognizing the difference between negative and meaningful conflict is important. Some conflicts are good things because they bring ideas and energy to a project.

As a project progresses, people conflicts will enter the picture. What causes conflicts? Team members may have multiple ideas on how something should be accomplished. Team members may be out of sync with each other. Team members may disagree on how to route a cable or where to start or how long it will take once things are started. When problems occur, there may not be agreement on the best way to solve the situation.

There are five options for resolving conflicts:

1. Withdrawing means that the project manager retreats from the disagreement. This is an option if the conflict is petty or of little impact on the project.

2. Smoothing is used to emphasize areas of agreement to help minimize or avoid areas of disagreement. This is the preferred method when people can identify areas of disagreement and the conflict is relatively unimportant.

3. Compromising involves creating a negotiated solution that brings some source of satisfaction to each party in the conflict. Compromises are best made after each side has had time to cool down. The best compromise makes each party feel as though he won.

4. Forcing is used when the boss exerts his or her position of power to resolve a conflict. This is usually done at the expense of someone else and is not recommended unless all other methods failed to resolve the conflict.

5. Confronting is not quite as strong as forcing, but it is the most common form of conflict resolution. The goal of a confrontation is to get people to face their conflicts directly, thereby resolving the problem by working through the issues in the spirit of problem solving.

Relevant Tools and Equipment

- *National Electrical Code® 2008*
- Calculator

Receptacles, Luminaires, and Switching Outlets (Student Manual) Lesson 428-5

Code References

- Article 100
- Article 210
- Article 680
- Article 410

Homework

1) Reading Assignment

- Read *Stallcup's Electrical Design Book*, Chapters 16 and 17

2) Key Terms

There are no new terms for this Lesson.

3) Practice Exercises

Answer the following referenced questions in the texts or other materials:

- In *Stallcup's Electrical Design Book* answer the odd numbered questions on pages 16-23, 16-24, 16-25, 17-23, 17-24, and 17-25.

Study Tips

▶ It is not too early to begin studying for your Journeyman's Exam. Many of the lessons you have studied this semester cover the same topics as lessons you have had in previous years. In preparing for your Journeyman's Exam, start by reviewing all your notes, homework, glossary of terms, and exams or quizzes for each lesson. Is there anything that you are still uncertain about? If so, be sure to ask your instructor about these areas during the class session. Or, if you prefer, ask your instructor for more material to review on these topics.

Receptacles, Luminaires, and Switching Outlets (Student Manual) Lesson 428-6

Tool Box Talk

Safety

✗ Working in hazardous environments increases the chances that an electrician may come into contact with hazardous workplace chemicals. Be on the lookout for such situations. Some chemicals may not be clearly marked. Although it is accepted practice to buy chemicals in bulk and repackage them into smaller containers for field use, the secondary containers may not be properly labeled even though they are required to be so. Be sure to check with the client before you handle, store, or transport chemicals in unlabeled containers.

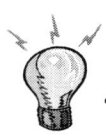

Tricks of the Trade

✓ With the exception of kitchen and bathrooms it is permissible to install a wall switch-controlled luminaire receptacle outlet instead of a fixed luminaire in habitable rooms.

✓ While most people prefer the convenience and light output of an overhead fixture, installation cost savings can be achieved by installing a switched luminaire receptacle outlet.

✓ A convenient way to differentiate which receptacle outlet in the room is the switched outlet is to install it upside down from the way you install all other outlets in the room. For example, if you normally install receptacle outlets with the ground down, install the switched outlet with the around up.

Plans, Specifications, and Documentation

When installing receptacle, lighting, and switching circuits, you should follow *NEC®* Article 410—Luminaires (Fixtures), Lampholders, Lamps, and Receptacles. However, these circuits involve more than one area of the *NEC®*. You should also review and be familiar with the following areas.

- Article 220—Branch Circuit, Feeder, and Service Calculations
- Article 240—Overcurrent Protection
- Article 250—Grounding and Bonding
- Article 300—Wiring Methods
- Article 310—Conductors for General Wiring
- Article 314—Outlet, Device, Pull and Function Boxes; Conduit Bodies, Fittings, and Hand Hole Enclosures
- Article 404—Switches

Motors and Compressor Motors (Student Manual) Lesson 429-1

I E C *PRIDE* **NATIONAL**	**Electrical Curriculum** **Year Four** **Student Manual**

Lesson 429 – Motors and Compressor Motors

Purpose

To review the *National Electrical Code®* requirements pertaining to motors and motor compressors in preparation for the Journeyman's Exam.

Objectives

By the end of this lesson, you should be able to:

429-1   Describe the five types of electric motors
429-2   Interpret and apply *NEC®* requirements for sizing motor branch and feeder circuit conductors
429-3   Interpret and apply *NEC®* requirements for sizing motor branch circuit protective devices
429-4   Interpret and apply *NEC®* requirements for sizing motor circuit overload and short circuit protective devices
429-5   Interpret and apply *NEC®* requirements for motor disconnecting means
429-6   Interpret and apply *NEC®* requirements for individual and group installations of hermetically sealed motor compressors

Content

Knowledge and Skills

- Motors
  - Branch—circuit and feeder circuit conductor sizing for:
  - Single motors
    - Singe-phase motors
    - Three-phase motors
    - Multispeed motors
    - Wye start and delta run motors

Motors and Compressor Motors (Student Manual) Lesson 429-2

- Duty cycle motors
- Adjustable speed drive systems
- Part-winding motors
- Wound-rotor secondary
- Continuous duty
- Other than continuous duty
- Resistors, separated from controller
- Several motors
- Duty cycle motors
- Motors and other loads

➢ Branch-circuit protective device
- Sizing per Table 430.251 and Table 430.252(B)
- Using instantaneous trip circuit breakers
- Motor type considerations
  - ☐ Single-phase AC squirrel-cage motors
  - ☐ Three-phase AC squirrel-cage motors
  - ☐ Wound-rotor motors
  - ☐ Synchronous
  - ☐ DC motors
- Motor code letters
  - ☐ Using code letters to find locked rotor current (LRC)
- Locked rotor current utilizing horsepower
- Sizing and selecting OCPDs
  - ☐ Nontime-delay fuses
  - ☐ Time delay fuses using maximum size
  - ☐ Instantaneous trip circuit breakers
  - ☐ Inverse time circuit breakers
  - ☐ Sizing maximum OCPD
  - ☐ Sizing OCPDs to allow motors to start and run
  - ☐ Nontime-delay fuses using the maximum size
  - ☐ Time-delay fuses using the maximum size
  - ☐ Inverse time circuit breakers
  - ☐ Several motors on one branch circuit
    - ■ Motors not over 1 hp
    - ■ Smallest rated motor protected
    - ■ Other group installations

➢ Sizing running overload protection
- Minimum size overload protection
- Maximum size overload protection

➢ Sizing the motor controller
- To start and stop the motor
- Stationary motor of 1/8 hp or less
- Portable motor of 1/3 hp or less
- Other than hp rated
- Design E motors
- Stationary motors
- Inverse time circuit breakers

Copyright © 2008 by the Independent Electrical Contractors, Inc. All rights reserved.

Motors and Compressor Motors (Student Manual) Lesson 429-3

- Torque motor
- Stationary and portable motors
➢ Motor disconnecting means
  - Sizing the disconnecting means
  - Other than hp rated
  - Design E motors
  - Stationary motors 1/8 hp or less
  - Stationary motors 2 hp or less
  - Motors 2 hp through 100 hp
  - Stationary motors
  - Cord-and-plug connected motors
  - Torque motor
  - Location of the disconnecting means
    - ☐ Within sight
      - ■ Locked in the open position
      - ■ Cannot be locked in the open position
➢ Motor control circuits
  - Sizing conductors for the control circuit
  - Protection of conductors
  - Control circuit transformers
  - Deenergized when the disconnecting means is open
  - Capacitor
  - Routing control circuit conductors
  - Occupying the same enclosure
  - Magnetic starter contactor and enclosure
➢ Determining motor full load current rating
➢ Number of disconnects served by each disconnect and controller
➢ Reduced starting methods
➢ Phase converters

- Compressor motors
  ➢ Nameplate information
  ➢ Markings on hermetic refrigerant motor-compressors and equipment
  ➢ Markings on controllers
  ➢ Ampacity rating highest rated(largest) motor
  ➢ Single machine
  ➢ Disconnecting means
  ➢ Rating and interrupting capacity
  ➢ Cord-connected equipment
  ➢ Location
  ➢ Branch-circuit short circuit and ground fault protection
    - Rating and setting for individual motor-compressor
    - Rating or setting for equipment
      - ☐ Sizing OCPD for two or more hermetic motors
      - ☐ Sizing OCPD for group protection
        - ■ Hermetic sealed motor is the largest load
        - ■ Hermetic sealed motor is not the largest load
    - Using 5 or 20 amp OCP

- Using a cord-and-plug connection not over 250 volts
- Single motor-compressors
- Two or more motor compressors
- Combination load
- Multimotor and combination load

➢ Controllers for motor compressors
- Motor compressor controller rating

➢ Motor compressor and branch-circuit overload protection
- Application and selection
- Overload relays
- On 15 or 20 amp branch circuits—not cord and plug connected
- On 15 or 20 amp branch circuits—cord and plug connected

➢ Room air conditioners
- Grounding
- Branch circuit requirements
- Disconnecting means
- Supply cords

Professionalism

- Ability to manage tasks, resources and costs effectively
- Ability to plan and organize work projects
- Ability to plan for time schedules
- Ability to use charts to schedule tasks
- Knowledge of budgets and budgeting
- Ability to use project control charts
- Knowledge of risk management
- Willingness to monitor progress on a project
- Willingness to take corrective action to ensure the success of a project
- Understanding the need for crisis management
- Ability to bring a project to completion

Motors and Compressor Motors (Student Manual) Lesson 429-5

Project Completion

The goal of project management is to obtain client acceptance of the project's end result. This means that the client agrees that the quality specifications were met. In order to have the acceptance stage go smoothly, the client and the project manager must have well-documented criteria for judging performance. These should be objective, measurable criteria that not subject to interpretation. There should be no room for doubt or ambiguity, although this is often difficult to achieve. It is also important to be clear what the project output is expected to accomplish.

The project may or may not be complete when results are delivered to the client. Often there are documentation requirements and a final report that still need to be provided. Finishing a project takes time and deliberate effort on your part. The final step of any project should be an evaluation review. This will look back over the project to see what was learned that will contribute to the success of future projects. This review is often best done as a team with those working on the project.

The following is a project completion checklist:

- Test project output to see that it works
- Write operations manual (if needed)
- Complete final drawings
- Train client's personnel to operate the project output
- Reassign project team
- Dispose of surplus equipment, material and supplies
- Summarize major problems encountered, solutions, and lessons learned
- Write performance evaluation reports
- Provide feedback on performance to all project staff
- Complete final audit and final report
- Conduct project review with upper management
- Declare project complete

Relevant Tools and Equipment

- *National Electrical Code® 2008*
- Calculator

Motors and Compressor Motors (Student Manual) Lesson 429-6

Code References

- Article 100
- Article 440
- 460.8
- Article 430

Homework

1) Reading Assignment

- Read *Stallcup's Electrical Design Book*, Chapters 18 and 19.

2) Key Terms

There are no new terms for this Lesson.

3) Practice Exercises

Answer the following referenced questions in the texts or other materials:

- In *Stallcup's Electrical Design Book* answer the odd numbered questions on pages 18-49, 18-50, 18-51, 18-52, 18-53, 18-54, 18-55, 19-17, 19-18, 19-19, 19-20, and 19-21.

Study Tips

▶ You should continue to study for your Journeyman's Exam at the same time you do the homework for this lesson. To help you remember the material, construct 1- or 2-page Study Sheets from your notes for each lesson. Use these sheets to summarize all the important information you will need to study for the exam. In so doing, you will force your mind to remember these topics, as well as help yourself by constructing review material you can go over right before taking the Journeyman's Exam.

▶ Last week's lesson was on receptacle outlets, switching, and luminary outlets. Review your homework. Try to list from memory all of the areas in which GFCI protection of receptacle outlets is required. Check yourself by reviewing *Stallcup's Electrical Design Book* and the *NEC®*.

Motors and Compressor Motors (Student Manual) Lesson 429-7

Tool Box Talk

Safety

✘ The first place to find out about a chemical is on its label, which will have warnings about any significant hazards from the material. Besides precautions, the label may recommend protective equipment and first-aid information. OSHA also regulates the proper labeling of chemicals.

Tricks of the Trade

✓ Remember that we frequently up-size an overcurrent protective device on a motor circuit to allow for starting current. The OCPD still provides protection against short circuits and ground faults and the addition of an overload block on the motor starter provides the necessary overcurrent protection.

✓ It is not uncommon on a motor circuit to see #12 conductors on a 50-amp breaker.

✓ Sometimes when we open an unfamiliar panelboard we may see what looks like improperly sized conductors on a breaker. Remember to first check to see if this is a motor circuit before assuming there is wiring mistake.

Plans, Specifications, and Documentation

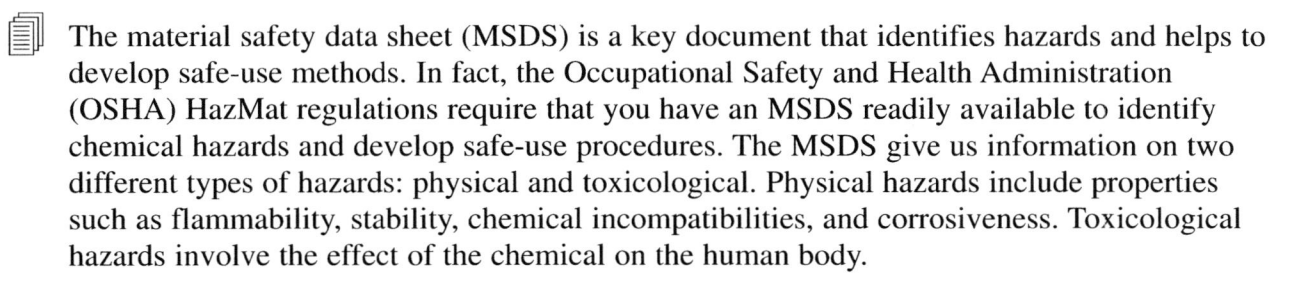

The material safety data sheet (MSDS) is a key document that identifies hazards and helps to develop safe-use methods. In fact, the Occupational Safety and Health Administration (OSHA) HazMat regulations require that you have an MSDS readily available to identify chemical hazards and develop safe-use procedures. The MSDS give us information on two different types of hazards: physical and toxicological. Physical hazards include properties such as flammability, stability, chemical incompatibilities, and corrosiveness. Toxicological hazards involve the effect of the chemical on the human body.

Hazardous Locations (Student Manual) Lesson 430-1

IEC NATIONAL PRIDE

Electrical Curriculum

Year Four Student Manual

Lesson 430 – Hazardous Locations

Purpose

This lesson will introduce you to hazardous locations as defined by the *National Electrical Code®*. In this lesson, you will learn about *NEC®* Articles 500 through 504. The lesson will familiarize you with hazardous location Classes I, II, and III. The lesson will also teach you about intrinsically safe systems.

Objectives

By the end of this lesson, you should be able to:
430-1   Describe the different types of hazardous locations
430-2   Discuss the *Code®* requirements for Class I installations
430-3   Discuss the *Code®* requirements for Class II installations
430-4   Discuss the *Code®* requirements for Class III installations
430-5   Define intrinsically safe locations
430-6   Discuss how to use *Code®* requirements governing intrinsically safe equipment

Content

Knowledge and Skills

- Describe the content of *NEC®* Article 500, Hazardous locations
  - Class I Groups
    - Group A
    - Group B
    - Group C
    - Group D
  - Class II Groups
    - Group E
    - Group F
    - Group G

- Equipment marking
- Class I temperatures
- Class II temperatures
- Class I locations
  - Division 1
  - Division 2
- Class II locations
  - Division 1
  - Division 2
- Class III locations
  - Division 1
  - Division 2

• Describe the contents of *NEC®* Article 501, Class 1 locations
- Transformers and capacitors
- Meters, instruments, and relays
- Wiring methods
- Sealing and drainage
- Switches, circuit breakers, motor controllers and fuses
- Motors and generators
- Lighting fixtures
- Utilization equipment
- Flexible cords
- Signaling, alarm, remote control, and communications systems
- Live parts
- Grounding
- Surge protection
- Multi-wire branch circuits

• Describe the contents of *NEC®* Article 502, Class 2 locations
- Transformers and capacitors
- Wiring methods
- Sealing
- Switches, circuit breakers, motor controllers and fuses
- Control transformers and resistors
- Motors and generators
- Ventilation piping
- Utilization equipment
- Flexible cords
- Receptacles and attachment plugs
- Signaling, alarm, remote control, and communications systems
- Grounding
- Surge protection
- Multi-wire branch circuits

• Describe the contents of *NEC®* Article 503, Class 3 locations
- Wiring methods
- Switches, circuit breakers, motor controllers and fuses
- Motors and generators

Hazardous Locations (Student Manual) Lesson 430-3

- Utilization equipment
- Lighting fixtures
- Flexible cords
- Receptacles and attachment plugs
- Signaling, alarm, remote control, and communications systems
- Cranes and hoists
- Storage battery charging equipment
- Grounding

• Describe the contents of *NEC®* Article 504, intrinsically safe systems
- Definition
- Equipment approval and installation
- Wiring methods
- Separation of conductors
- Grounding
- Bonding

Professionalism

- Ability to identify the different Classes, Groups, and Divisions of hazardous locations
- Comfort in the use and interpretation of the codes

Communication Skills

Communication is an important part of your job, including working with your team and with customers. Here are some keys to talking so others will listen.

- **Determine why you are speaking.** What is the purpose? What you talking about? Why is it important?

- **Know your customer or audience.** Who will listen to your message? What is important to this person or group? How will the audience feel about your message?

- **Prepare your message.** Gather information and plan your key points. Be brief, be clear, and have the facts to back up what you say. After you tell someone something, pause. If you don't pause, they can't give you feedback or ask questions. Practice pausing for three to four seconds and see how much more often people interact with you.

- **Check your spoken image and your body image.** Work to eliminate bad habits such as rocking, saying "uh" or looking down while speaking or listening. Your body language can speak as "loudly" as the words you say.

Practice, Practice, Practice. As a professional, you may have to deliver news to your customer that they don't want to hear! Being prepared is key to doing this well.

Hazardous Locations (Student Manual) Lesson 430-4

Relevant Tools and Equipment

- *National Electrical Code® 2008*
- Calculator

Code References

- Article 500
- Article 501
- Article 502
- Article 503
- Article 504

Homework

1) Reading Assignment

- Read 2008 *National Electrical Code®* Articles 500, 501, 502, 503, and 504

2) Key Terms

Using either the *Illustrated Dictionary for Electrical Workers*, the Glossary in the Appendix or your textbooks, write definitions for each of these terms before doing your homework:

- Class I location
- Class II location
- Class III location
- Combustible
- Division 1
- Division 2
- Dust ignition proof
- Explosion proof
- Group A
- Group B
- Group C
- Group D
- Group E
- Group F
- Ignitable
- Intrinsically safe
- Intrinsically safe apparatus
- Intrinsically safe circuit
- Intrinsically safe system

Hazardous Locations (Student Manual) Lesson 430-5

3) Practice Exercises

- None.

Study Tips

▶ There are many Classes and Division locations to learn in this lesson and so it would help you to develop your own chart that shows the hazardous locations, their characteristics, and *Code®* Articles. It is important that you design and construct this chart yourself. By going through this process, you will be able to remember the material.

▶ You will learn a lot of new concepts in this lesson. The following are some memory tips that can help you to remember all these terms.

- If you want to remember something, review it immediately, and then again within 20 minutes.
- Use a spaced practice schedule—20 minutes a day for five days is much better than two hours of memorization at one sitting.
- Make the information meaningful. Don't try to memorize information that you don't understand.
- Organize the information into chunks of seven items or less. If you have to memorize a big list of something, group into smaller lists organized in some logical fashion—even if only you know that logic.

Tool Box Talk

Safety

✗ A type of hazard warning that you often see on bulk containers is the HMIG (hazardous material identification Manual) label, which has four boxes—blue, red, yellow, and white—each with a number from 1 to 4 in them. Each color identifies a type of hazard with the number identifying the severity. The blue is health, red is flammability, yellow is reactivity (explosiveness), and white contains symbols for recommended protective equipment. If you are using any chemical with a rating of "2" or above for the blue, red, or yellow colors, you should be familiar with the details of the particular hazard that causes the rating.

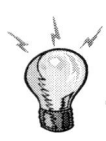

Tricks of the Trade

✓ Don't be confused by the term "explosion proof." When we are wiring in hazardous locations we use either intrinsically safe methods or explosion proof conduit systems.

✓ The term explosion proof leads many to think that there could never be any ignition of a hazardous atmosphere but if the hazardous atmosphere gets into the conduit a spark from switching could ignite the gaseous mixture in the conduit system. The explosion proof conduit system will contain the ignition and prevent the outside atmosphere from igniting as well.

Plans, Specifications, and Documentation

Material safety data sheets (MSDS) identify two types of hazards. **Physical hazards** involve how we handle the chemical and the surrounding environment. Warnings like "Do not use near fire and flame" or "Do not store in metal container" relate to physical hazards. **Toxicological hazards** relate to our personal contact with the chemical. Warnings like "Wash hands before eating or smoking," "Use with adequate ventilation," or "Wear impervious gloves to prevent skin contact," relate to toxicological hazards.

Special Types of Hazardous Locations (Student Manual) Lesson 431-1

| *I E C*
 PRIDE
 NATIONAL | Electrical Curriculum

 Year Four
 Student Manual |

Lesson 431 – Special Types of Hazardous Locations

Purpose

This lesson will introduce you to a variety of hazardous locations covered by the *National Electrical Code®*. In this lesson, you will learn about *NEC®* Articles 511 through 516. These Articles give *NEC®* requirements for special types of hazardous location installations, such as commercial garages, aircraft hangers, gasoline stations, bulk storage, and painting, dipping, and coating installations.

Objectives

By the end of this lesson, you should be able to:

431-1   Describe the *Code®* requirements for commercial garage installations
431-2   Describe the *Code®* requirements for aircraft hanger installations
431-3   Describe the *Code®* requirements for gasoline station installations
431-4   Describe the *Code®* requirements for bulk storage plant installations
431-5   Describe the *Code®* requirements for spray painting, dipping, and coating room installations

Content

Knowledge and Skills

- Describe the contents of *NEC®* Article 511, Commercial garages
  ➢ Class I locations
    – Wiring and equipment
    – Sealing
  ➢ Wiring and equipment above Class I locations
  ➢ Battery and vehicle charging
  ➢ GFCI protection for personnel

Copyright © 2008 by the Independent Electrical Contractors, Inc. All rights reserved.

- Describe the contents of *NEC®* Article 513, Aircraft hangers
  - Classifications of locations
  - Wiring and equipment in Class I locations
  - Stanchions, rostrums, and docks
  - Sealing
  - Aircraft electrical systems
  - Battery charging and equipment
  - External aircraft power sources
  - Mobile servicing equipment

- Describe the contents of *NEC®* Article 514, Gasoline dispensing and service stations
  - Class I locations
  - Wiring and equipment within Class I locations
  - Wiring and equipment above Class I locations
  - Circuit disconnects
  - Sealing
  - Table 514.3 (B) (1)
  - Table 514.3 (B) (2)

- Describe the contents of *NEC®* Article 515, Bulk storage plants
  - Class I locations
  - Wiring and equipment in Class I locations
  - Wiring and equipment above Class I locations
  - Underground wiring
  - Table 515.3
  - Sealing
  - Gasoline dispensing
  - Grounding

- Describe the contents of *NEC®* Article 516, Spray application, dipping, and coating processes
  - Classifications of locations
  - Wiring and equipment in Class I locations
  - Electrostatic hand spraying equipment
  - Powder coating
  - Wiring and equipment above Class I and Class II locations

Professionalism

- Ability to work in a team environment

Special Types of Hazardous Locations (Student Manual) Lesson 431-3

Team Chemistry

Teamwork requires that all of you pool information and consider different viewpoints to find solutions and make decisions. Yet, the chances are pretty slim that you will always agree on every point, if you let the loudest person make the decisions or just go by majority rule, you may ignore valuable input from your quieter teammates.

What happens to your motivation when your ideas are ignored or rejected? How invested do you feel if your suggestions aren't even considered? If these things happen, teamwork may suffer and eventually collapse.

Teams will not bond quickly or solidly without team chemistry. Face it: there is more to teamwork that just facts, plans and work processes. Because the human element is vital to your success, you need to routinely make time to work on team building. This can involve having regular, but short meetings to get everyone's input on a job or project. This can involve having fun together. This can involve taking time to resolve differences. Mostly, this involves taking time to get to know each other better.

Relevant Tools and Equipment

- *National Electrical Code® 2008*
- Calculator
- Blueprints for one or more types of special hazardous installations

Code References

- Article 250
- Article 511
- Article 513
- Article 514
- Article 515
- Article 516

Homework

1) Reading Assignment

- Read *National Electrical Code®* Articles 511, 513, 514, 515, and 516.

2) Key Terms

Write definitions for each of the key terms before you do your homework. Then go back and make any corrections to your understanding of the key terms after you have studied the homework assignment.

- Aircraft hanger
- Bulk storage plant
- Commercial garage
- Dispenser
- Pit

3) Practice Exercises

- None

Study Tips

▶ This lesson deals with identifying and distinguishing among various *Code®* requirements. You will need to memorize a lot of new material. To do this, you will need to work on understanding the organization behind the different Classes and Divisions in the *Code®*. The more time that you spend on understanding this organization, the more likely you are to remember the material.

▶ One good way to review for the Journeyman's exam is to review the objectives for each lesson, particularly the lessons for Years 3 and 4. Ask yourself, "Can I do the things that are listed in the objectives?" If you can, you probably have a good understanding of the material and only need a quick review. However, if you find lessons where you can't measure up to the objectives, you should focus on those lessons for careful study and review.

Special Types of Hazardous Locations (Student Manual) Lesson 431-5

Tool Box Talk

Safety

✗ When using any chemicals, safety is a major concern. Many chemicals may be harmful if inhaled, ingested, or exposed to skin; some are flammable and may react with other substances. **Isopropyl alcohol**—also called isopropanol—is a common chemical you may encounter. It comes in a variety of forms: a straight solution within a container, premoistened lint-free wipes, foam swabs that house the alcohol within a capillary that comprises the swab handle, and a canned aerosol spray. The 99% alcohol is highly combustible, volatile, and flammable. Improperly protected skin that is exposed to 99% alcohol will quickly dry out and become irritated. Inhalation of the vapors and overexposure to alcohol should be avoided at all costs, since prolonged exposure could result in kidney failure.

Tricks of the Trade

✓ In this lesson we explored some special occupancies that we've never discussed before in this program. The *NEC®* has many sections devoted to Special Occupancies in Chapter 5 and Special Equipment in Chapter 6.

✓ Depending on where you end up in your career you may see some of these frequently, but some of them you may never see at all. There is no way you can memorize everything about all these occupancies and types of equipment. That is why it is important to be able to use the *Code®* as a tool to help you reference the information when you need it.

✓ You are not finished with the *Code®* book when you finish your apprenticeship program. It is a tool you will use throughout your career.

Plans, Specifications, and Documentation

To remember the different kinds of locations discussed by the *NEC®*, keep the basis for the classification scheme in mind. Locations are classified depending on the properties of the flammable vapors, liquids, gases, or combustible dusts or fibers that are present. Locations are also classified based on the likelihood that a flammable or combustible concentration or quantity is present. Each room, section, or area of a building must be considered individually. However, even if a particular room is "unclassified," it may be next to a location where a hazard exists and so it may also be considered a hazardous location. As you go through the *NEC®*, make sure you note the essential condition(s) or situation.

Copyright © 2008 by the Independent Electrical Contractors, Inc. All rights reserved.

IEC *PRIDE* **NATIONAL**

Electrical Curriculum

Year Four
Student Manual

Lesson 432 – Signs and Sign Connections

Purpose

This lesson covers the installation of conductors and equipment for electric signs and outline lighting as addressed in the *National Electrical Code®* Article 600. Article 600 pertains to electric signs of a fixed, stationary, or portable self-contained type. In this lesson, you will learn about the general requirements for such lighting in regard to branch circuits, disconnects, grounding, and other important features, such as transformers, ballasts, electronic power supplies, and neon tubing.

Objectives

By the end of this lesson, you should be able to:

432-1   Interpret and apply *NEC®* requirements for signs
432-2   Interpret and apply *NEC®* requirements for field installed skeleton tubing

Content

Knowledge and Skills

- Branch circuits
- Disconnects
- Grounding
- Portable or mobile signs
- Ballasts, transformers and electronic power supplies
- Neon secondary circuit conductors, 1000 volts or less nominal
- Neon secondary circuit conductors over 1000 volts, nominal
- Neon tubing
- Electrode connections

Signs and Sign Connections (Student Manual) Lesson 432-2

Professionalism

- Learn to use vendor specifications
- Practice use of application manuals and installation practices manuals for products

Read and Apply Technical Knowledge: Reading Vendor Specifications and Directions.

Your textbooks are not the only place that you can use to learn important technical information. In fact, this field is changing so fast that information may change before book publishers can issue new versions of textbooks. One of the best places to get up-to-date technical information is from vendor catalogs, instruction and specification sheets. Vendors will have the most accurate and up-to-date information on both the performance specifications of their components and the proper steps for installation of these components. These instructions help to assist in identifying the unique features of each component and point out how this component can be used for a reliable and trouble-free installation. They often provide safety tips and tricks of the trade. They include phone numbers and web sites that can put you in touch with experienced technical staff that can help you with any questions or problems.

When you open the container with new components, don't just throw away instruction and specification sheets, assuming that you have seen them before. Take the time to review these sheets, looking for new information and reminding yourself of tips that will help you to do installations more quickly and safely. Keep up-to-date catalogs for the vendor of each major product that you use on the job. Take time to review these catalogs. At the very least, they will give you information about new products that you may need to use on the job. Many of them provide sections including Application Manuals, Installation Practices, Glossaries of Terms and, even, Overviews of Standards. In a field that changes as rapidly as this one, these resources provide you with the latest information and keep you informed of changes in the field.

Relevant Tools and Equipment

- *National Electrical Code® 2008*
- Calculator

Code References

- Article 600

Signs and Sign Connections (Student Manual) Lesson 432-3

Homework

1) Reading Assignment

- Read *NEC®* Article 600.

2) Key Terms

Write definitions for each of the key terms before you do your homework. Then go back and make any corrections to your understanding of the key terms after you have studied the homework assignment.

- Electric-discharge lighting
- Neon tubing
- Sign body
- Skeleton tubing

3) Practice Exercises

Answer the following referenced questions in the texts or other materials:

- None

Study Tips

▶ In order to better understand and remember the *Code®* requirements in this lesson, try to visualize yourself using Article 600 to install sign circuits. How would you proceed? What are the most important parts of this Article? What are the important exceptions? Are there parts that are more confusing than others? All of these are the kinds of questions that you should answer to help you remember the information in this Article.

Tool Box Talk

Safety

✗ When using any chemicals, safety is a major concern. Many chemicals may be harmful if inhaled, ingested, or exposed to skin; some are flammable and may react with other substances. **Solvents or cleaners** are another commonly used type of chemical. These solutions—either a petroleum- or water-based product—are available in a container, a spray bottle, or pre-moistened wipes. They also remove grease, tar, oils, and dirt from hands, cables, tools, and equipment, so they are considered a relatively safe, multipurpose chemical. However, prolonged exposure can make you ill and you need to take precautions when using these chemicals.

Tricks of the Trade

✓ Neon signs were very popular in the late 1950s and throughout the 1960s. This form of outline luminaire fell out of favor for a while but now it making a comeback.

✓ In the 50s and 60s, all Neon signage was custom fabricated and required a close collaboration between the electrician and the glassblower.

✓ Many modern signs are now being fabricated entirely in the glass shop using off the shelf transformers designed for the glass industry but large custom jobs still require a close working relationship between the electrician and the glass blower.

✓ Some custom signs take a considerable period of time to manufacture. You will want to start working with the glassblower early in the project to insure all the parts of the project come together at the appropriate time.

✓ Did you know? The type of gas used in the tubing determines the color displayed when the electric arc passes through it?

Plans, Specifications, and Documentation

Article 110.2 tells us conductors and equipment required or permitted by the *Code®* shall be acceptable only if approved and approved is further defined as acceptable to the authority having jurisdiction. This is usually considered to mean that an item must be listed by a nationally recognized testing laboratory. It is important when working with neon designers and glassblowers to insure any electrical components they may be furnishing are listed before we install them on our job.

Load Calculations (Student Manual) Lesson 433-1

| *I E C* **PRIDE** *NATIONAL* | **Electrical Curriculum**

Year Four
Student Manual |

Lesson 433 – Load Calculations

Purpose

To review the techniques for calculating branch circuit, feeder, and service loads for residential, commercial, and industrial occupancies.

Objectives

By the end of this lesson, you should be able to:

433-1   Perform load calculations for single and multifamily residences
433-2   Perform load calculations for commercial occupancies
433-3   Perform load calculations for industrial facilities

Content

Knowledge and Skills

- Residential Calculations
  - Applying the standard calculation for single family dwelling units (*NEC®* Article 220, Part II)
    - General luminaire load
    - Small appliance and laundry load
    - Special appliance loads
    - Demand factors
      - ☐ General luminary and receptacle, small appliance, and laundry loads
      - ☐ Cooking equipment loads
      - ☐ Fixed appliance load
      - ☐ Dryer load
    - Largest load between heat and a/c
    - Largest motor load

Copyright © 2008 by the Independent Electrical Contractors, Inc. All rights reserved.

- Applying the optional calculation for single family dwelling units
  - New construction (*NEC®* 220.80, Part IV)
    - ☐ Other loads
    - ☐ Heat or A/C loads
  - Existing units
    - ☐ Other loads
    - ☐ Added appliance load
- Multifamily dwellings
  - Standard calculation
  - Optional calculation
- Feeder to mobile home—standard calculation
- Mobile home park service and feeders—optional calculation
- Other calculations
  - Sizing the service conductors
  - Sizing the service OCPD
  - Sizing the panelboard
  - Sizing the rigid metal conduit for the service conductors
  - Sizing the grounding electrode conductor (GEC) to ground the service to a metal water pipe
  - Sizing the grounding electrode conductor (GEC) to ground the service to a driven rod
  - Sizing the grounded conductor to clear a ground fault

- Commercial Calculations
  - Applying the standard calculation
    - Luminary loads
      - ☐ General luminaire loads
      - ☐ Noncontinuous and continuous operation
      - ☐ Listed occupancies
      - ☐ Unlisted occupancy
      - ☐ Show window luminaire load
      - ☐ Track luminaire load
      - ☐ Low-voltage luminaire load
      - ☐ Outside luminaire load
      - ☐ Outside sign luminaire load
    - Receptacle loads
      - ☐ Applying demand factors
      - ☐ Multioutlet assemblies
    - Special appliance loads
      - ☐ Continuous and noncontinuous operation
      - ☐ Applying demand factors
    - Compressor loads
      - ☐ Continuous or noncontinuous operation
    - Motor loads
    - Heat or A/C loads
    - Largest motor load
  - Applying the optional calculation
    - Kitchen equipment
    - Schools
    - Restaurants

Load Calculations (Student Manual) Lesson 433-3

- ➢ Optional calculations for additional loads to existing installation
  - When demand data is not available
- Industrial Calculations
  - ➢ Standard Calculations
    - Luminaire loads
      - ☐ Unlisted occupancy
      - ☐ Show window luminaire
      - ☐ Track luminaire
      - ☐ Low voltage luminaire
      - ☐ Outside luminaire
      - ☐ Outside sign
    - Receptacle loads
      - ☐ General-purpose receptacles
      - ☐ Applying demand factors
      - ☐ Multioutlet assemblies
    - Special loads
      - ☐ Continuous and noncontinuous operation
    - Motor loads
    - Largest between heat and A/C loads
    - The largest motor load
  - ➢ Optional Calculations for additional loads to existing installations
    - When demand data is available
    - When demand data is not available
  - ➢ Computing the neutral
  - ➢ Using one-line diagrams

Professionalism

- Ability to communicate effectively
- Ability to avoid barriers that keep you from being understood
- Understand why misunderstandings occur
- Demonstrate how to minimize or avoid misunderstandings

Communicating Effectively in a Team Environment: Eliminating the Negatives

Effective communications between two people exists when one understands the message the way the other person intended. A communication gap exists when one person reacts to a message in a manner entirely different from what was intended. One of the first major steps to achieve more effective communication between people is to recognize and understand why misunderstandings occur in the first place and then learn how to minimize or avoid them. We will talk about two aspects of communicating effectively. In this lesson, we will talk about the barriers to better communication.

- **Hearing what you expect to hear.** As human beings, we tend to listen for only those things that we expect to hear and we screen out everything else.

- **Evaluating the source.** Human beings tend to evaluate the source of the information they receive to determine its value.

- **Having different perceptions.** If you are upset with the way a co-worker communicates with you, you probably assume the co-worker wanted you to be upset.

- **Non-verbal communications.** Not all non-verbal signals are accurately interpreted. For example, most books on non-verbal attitudes will tell you that people who cross their arms and legs while listening are "tuning out." Not necessarily so! Many times the listener is displaying this behavior in order to concentrate better.

- **Being distracted by noise.** Usually it is more difficult to concentrate and communicate effectively when there's a lot of noise around us. To be able to better send and receive messages, it's preferable to find a quiet place to conduct business.

Relevant Tools and Equipment

- *National Electrical Code® 2008*
- Calculator

Load Calculations (Student Manual) Lesson 433-5

👉 Code References

- Article 110
- Article 210
- Article 220
- Article 250
- Article 408
- Article 430
- Article 550
- Article 200
- Section 215.2
- Section 230.42
- Section 310.15
- Article 440

Homework

📖 1) Reading Assignment

- Read *Stallcup's Electrical Design Book*, Chapters 22, 23, and 24.

AA 2) Key Terms

There are no new terms for this Lesson.

3) Practice Exercises

Answer the following referenced questions in the texts or other materials:

- In *Stallcup's Electrical Design Book* answer the odd numbered questions on pages 22-29 through 22-33. Answer the odd number questions on pages 23-35 through 23-41. Answer the odd numbered questions on page 24-15.

Study Tips

▶ You should continue to study for your Journeyman's exam at the same time you do the homework for this lesson. Concentrate on the least familiar areas. Don't spend as much time studying those areas you use in your job every day unless there is an area in which you are particularly weak.

Safety

✗ Always do a quick safety check to identify possible safety hazards. Each electrician is responsible for identifying safety hazards and taking proper precautions. "An ounce of prevention is worth a pound of cure." When you get busy wiring, you may be too busy to be aware of these things.

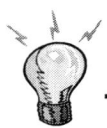

Tricks of the Trade

✓ When performing a service calculation for a large commercial or industrial structure, remember to allow for the different types of occupancies when calculating the luminaire load portion. As an example, the VA per square foot for an office building according to Table 220.12 is 3 1/2. But if we remember that building may have a restaurant at 2 VA per square foot, an auditorium at 1 VA per square foot, hallways at 1/2 VA per square foot and storage spaces at 1/4 VA per square foot. We can reduce the overall luminaire load requirement from what we would have had if we had calculated the entire building at 3 1/2 VA allowing us to come up with a smaller total load and possibly reduce a feeder size.

Plans, Specifications, and Documentation

Just because an electrical installation conforms to the safety requirements of the *National Electrical Code®* does not eliminate sloppy workmanship or improper design. *NEC®* Article 90-1 (b) states that "This *Code®* contains provisions considered necessary for safety. Compliance therewith and proper maintenance will result in an installation essentially free from hazard but not necessarily efficient, convenient, or adequate for good service or future expansion of electrical use." While safety should always be a primary consideration, electricians should never take the adequacy of an electrical system on face value because it meets *NEC®* requirements for safety.

Final Code Review and Test Preparation (Student Manual) Lesson 434-1

IEC NATIONAL PRIDE

Electrical Curriculum

Year Four
Student Manual

Lesson 434 – Final Code Review and Test Preparation

Purpose

This lesson will give students an opportunity to review any *National Electrical Code®* requirements that the students still may not understand. The lesson will also review local codes and test peculiarities to prepare the students for the Journeyman's Exam.

Objectives

By the end of this lesson, you should be able to:

434-1   Interpret and apply *NEC®* requirements
434-2   Interpret and apply requirements of local codes pertaining to electrical work
434-3   Take and pass the state and/or local Journeyman's Exam

Content

Knowledge and Skills

- *National Electrical Code® 2008*
- Any pertinent state or local codes

Professionalism

- Ability to ask questions or to ask for clarification when something is unclear
- Ability to plan and organize learning materials
- Willingness to work with others to solve problems as a team

Copyright © 2008 by the Independent Electrical Contractors, Inc. All rights reserved.

Final Code Review and Test Preparation (Student Manual) Lesson 434-2

Consensus Problem-Solving

Troubleshooting problems with electrical installation is a form of problem solving. Troubleshooting or problem solving is an investigative process that will often require us to work as a team. The team must agree on the solution to the problem, otherwise we will work at cross-purposes and waste time and energy. Team consensus is a decision or position that reflects the thinking of the team and that all team members participate in developing and actively supporting. Team consensus is an idea that merges the best thinking of all team members. To reach consensus, all team members must:

- Express themselves clearly and participate fully
- Be open-minded, listen fully and respect others' views
- Acknowledge others contributions and provide feedback on their ideas
- Propose solutions to differences and be willing to negotiate
- Identify areas of agreement and work from there to seek consensus

In the next lesson, we will talk about techniques or "tools" that we can use to do team problem solving more quickly and effectively.

Relevant Tools and Equipment

- *National Electrical Code® 2008*
- Calculator

Code References

- *National Electrical Code® 2008*
- Any pertinent local codes

Homework

1) Reading Assignment

- In the 2008 *National Electrical Code®*, review Articles 90, 110, and the examples given in Annex D, pp. 717–725.

Final Code Review and Test Preparation (Student Manual) Lesson 434-3

2) Key Terms

Review all of the terms in Article 100 of the *National Electrical Code®*. Make sure that you understand all of them.

3) Practice Exercises

Answer the following referenced questions in the texts or other materials:

- None

Study Tips

Here are some steps that may help you to study for the Journeyman's Exam.

▶ Step one: Decide which areas are your weakest and devote more time and energy to studying those areas.

▶ Step two: Learn as much as possible about what to expect on the exam. Ask your instructor and electricians whom you work with what to expect on the exam.

▶ Step three: Decide on a study schedule that will help you to cover all areas for review. Plan carefully so that you don't neglect any important areas in your review.

▶ Step four: Work with a study group. Take turn quizzing each other. Give each other sample problems to solve. Review terms with each other.

▶ Step five: If you have time, go over all the homework from each lesson. If you don't have that much time, focus on those homework lessons that gave you the most trouble.

ESTIMATING SYSTEMS
ESTIMATING CENTER

MCCORMICK
McCormick Systems, Inc.

Proudly Supporting Over 34,000 Systems Since 1979

Remote Takeoff

Win T-Bill

CAD Estimating

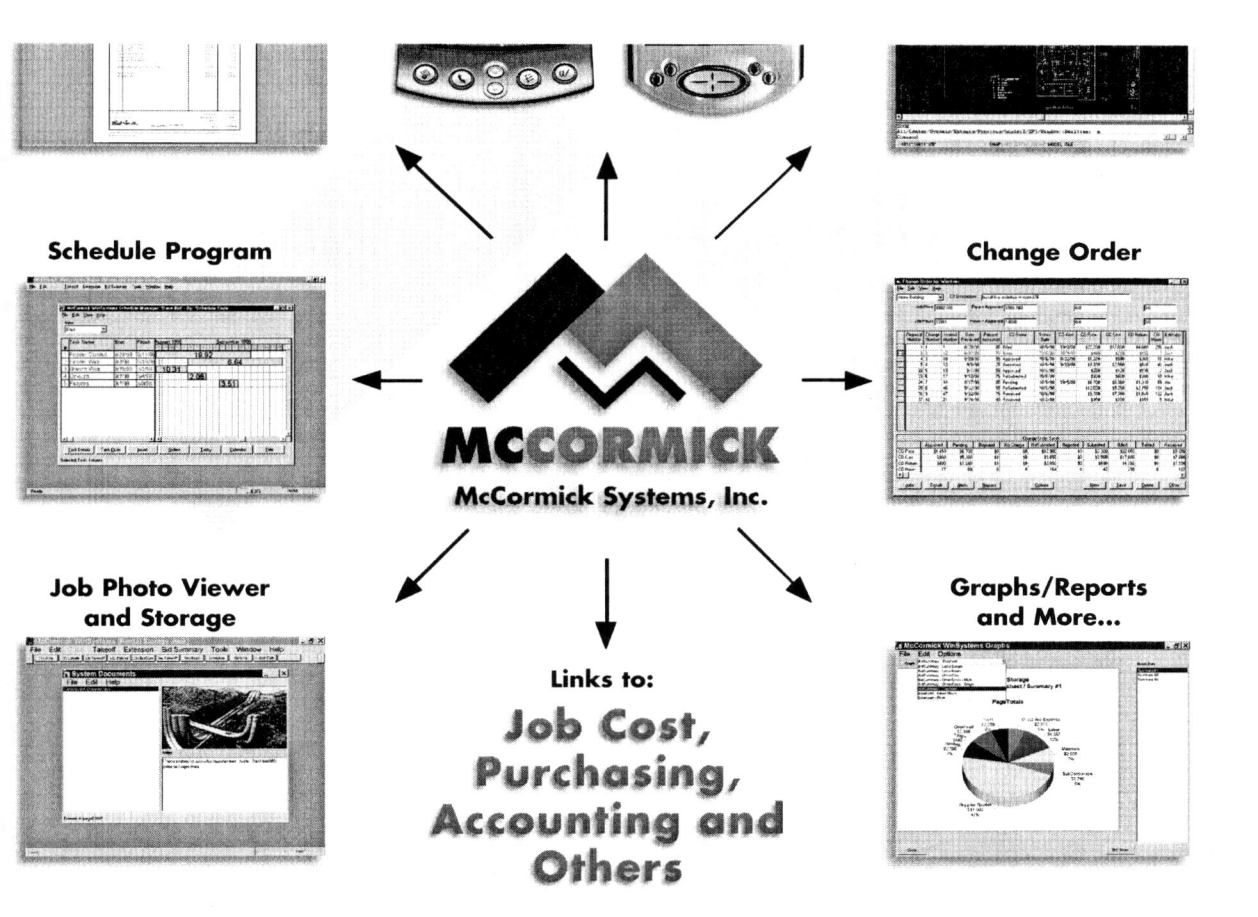

Schedule Program

Change Order

McCormick
McCormick Systems, Inc.

Job Photo Viewer and Storage

Links to:
Job Cost, Purchasing, Accounting and Others

Graphs/Reports and More...

The Nations Leader in Estimating Software!

CALL FOR MORE INFORMATION AND VISIT OUR WEB SITE!
1-800-444-4890 • 480/831-8914 • www.mccormicksys.com

Final Code Review and Test Preparation (Student Manual) Lesson 435-1

IEC PRIDE NATIONAL

Electrical Curriculum

Year Four
Student Manual

Lesson 435 – Final Code Review and Test Preparation

Purpose

To review Lessons 419 through 434. Prepare students to take the Final Exam that has been compiled.

Objectives

By the end of this lesson, you should be able to:

419-1 Describe major power quality problems and sources
419-2 Discuss the basic concepts and procedures for power quality measurement
419-3 Discuss the effects of poor power quality on electrical equipment and systems
419-4 Discuss indicators of power quality problems and the equipment and methods used to resolve these problems

420-1 Summarize the various codes and standards that will impact you as you pursue your trade as an electrician
40-2   Summarize the various types of injuries that can result from electrical shock and discuss treatment for each
420-3 Identify and use appropriate electrical protective clothing
420-4 Describe the various types of voltage systems typically used in residential, commercial, and industrial applications
420-5 Interpret any *Code®* requirements related to low voltage systems
420-6 Identify any *Code®* requirements related to working clearances

421-1 Identify the major and minor ways in which electricity is produced and distributed
421-2 Connect wye and delta transformer bank connections
421-3 Describe the functions of a substation
421-4 Explain the difference between feeders, busways and other downstream systems
421-5 Discuss motor control centers and systems

Final Code Review and Test Preparation (Student Manual) Lesson 435-2

422-1	Interpret *Code®* requirements related to services
422-2	Interpret *Code®* requirements related to switchboards and panelboards
423-1	Identify conductor types and their intended application
423-2	Determine the ampacity of conductors
423-3	Compute proper circuit loading
423-4	Match temperature markings
423-5	Interpret and apply *NEC®* requirements for selecting and sizing conductors
423-6	Interpret and apply *NEC®* requirements for derating conductors
423-7	Interpret and apply *NEC®* requirements for protection of equipment conductors, and tap conductors
423-8	Select and apply circuit breakers and fuses in accordance with *NEC®* requirements
423-9	Interpret and apply *NEC®* requirements for ground fault protection for equipment and personnel
424-1	Interpret and apply *NEC®* requirements for circuit and system grounding
424-2	Interpret and apply *NEC®* requirements for equipment grounding
424-3	Interpret and apply *NEC®* requirements for supply and load equipment bonding
424-4	Interpret and apply *NEC®* requirements for grounding electrode systems
425-1	Interpret and apply *NEC®* requirements for sizing boxes, conduit and fittings based on required conductor and device fill
425-2	Interpret and apply *NEC®* requirements for conductor fill for gutter, auxiliary gutters, panelboards, and cable tray
425-3	Interpret and apply *NEC®* requirements for box support
425-4	Interpret and apply *NEC®* requirements for AC, MC, NM, and SE cable
425-5	Interpret and apply *NEC®* requirements for RMC, IMC, and EMT
425-6	Interpret and apply *NEC®* requirements for flexible metal conduit and liquid tight flexible metal conduit
425-7	Interpret and apply *NEC®* requirements for the installation of cable and raceway systems
426-1	Interpret and apply *NEC®* requirements for residential branch circuits
426-2	Interpret and apply *NEC®* requirements for commercial and industrial branch circuits
426-3	Interpret and apply *NEC®* requirements for feeder circuits
426-4	Interpret and apply *NEC®* requirements for AC, MC, NM, and SE cable
426-5	Interpret and apply *NEC®* requirements for RMC, IMC, and EMT
426-6	Interpret and apply *NEC®* requirements for flexible metal conduit and liquid tight flexible metal conduit
426-7	Interpret and apply *NEC®* requirements for the installation of cable and raceway systems
428-1	Interpret and apply *NEC®* requirements for residential receptacles outlets
428-2	Interpret and apply *NEC®* requirements for GFCI protection of residential receptacle outlets
428-3	Interpret and apply *NEC®* requirements for GFCI protection of receptacles in and around boathouses, swimming pools, storage pools, spas, hot tubs, and hydromassage tubs
428-4	Interpret and apply *NEC®* requirements for receptacles on construction sites

Final Code Review and Test Preparation (Student Manual) Lesson 435-3

428-5	Interpret and apply *NEC*® requirements for receptacles installed in commercial and industrial locations
428-6	Interpret and apply *NEC*® requirements for residential luminaire and switching outlets
428-7	Interpret and apply *NEC*® requirements for commercial and industrial luminaire and switching outlets
428-8	Interpret and apply *NEC*® requirements for switching outlets by pools, hot tubs, and spas
429-1	Describe the five types of electric motors
429-2	Interpret and apply *NEC*® requirements for sizing motor branch and feeder circuit conductors
429-3	Interpret and apply *NEC*® requirements for sizing motor branch circuit protective devices
429-4	Interpret and apply *NEC*® requirements for sizing motor circuit overload and short circuit protective devices
429-5	Interpret and apply *NEC*® requirements for motor disconnecting means
429-6	Interpret and apply *NEC*® requirements for individual and group installations of hermetically sealed motor compressors
430-1	Describe the different types of hazardous locations
430-2	Discuss the *Code*® requirements for Class I installations
430-3	Discuss the *Code*® requirements for Class II installations
430-4	Discuss the *Code*® requirements for Class III installations
430-5	Define intrinsically safe locations
430-6	Discuss how to use *Code*® requirements governing intrinsically safe equipment
431-1	Discuss the *Code*® requirements for commercial garage installations
431-2	Discuss the *Code*® requirements for aircraft hanger installations
431-3	Discuss the *Code*® requirements for gasoline station installations
431-4	Discuss the *Code*® requirements for bulk storage plant installations
431-5	Discuss the *Code*® requirements for spray painting, dipping, and coating room installations
432-1	Interpret and apply *NEC*® requirements for signs
432-2	Interpret and apply *NEC*® requirements for field-installed skeleton tubing
433-1	Perform load calculations for single and multifamily residences
433-2	Perform load calculations for commercial occupancies
433-3	Perform load calculations for industrial facilities
434-1	Interpret and apply *NEC*® requirements
434-2	Interpret requirements of local codes pertaining to electrical work
434-3	Take and pass the state and/or local Journeyman's Exam

Content

Knowledge and Skills

- Fire alarm systems
- *Code®* requirements pertaining to the following areas:
  - Electrical Code and Standards
  - Electrical Safety and First Aid
  - Electrical Systems
  - Working Clearances
  - Services
  - Switchboards and Panelboards
  - Conductors
  - Overcurrent Protection Devices
  - Lightning Protection
  - Grounding
  - Designing Wiring Methods
  - Installing Wiring Methods
  - Branch-circuits
  - Feeder-circuits
  - Receptacle Outlets
  - Luminaires and Switching Outlets
  - Motors
  - Compressor Motors
  - Transformers
  - Hazardous (Classified) Locations
  - Signs
  - Residential Calculations
  - Commercial Calculations
  - Industrial Calculations
- Local code requirements
- Preparations for the Journeyman's Exam

Professionalism

- Willingness to memorize and use new terms
- Ability to ask questions or to ask for clarification when something is unclear
- Ability to listen to a large amount of information and then organize it for recall.
- Ability to work with others to solve problems

Final Code Review and Test Preparation (Student Manual) Lesson 435-5

Techniques for Group Problem Solving

The question that often comes up when groups try to solve problems is how do we do this quickly and effectively. There are three phases of communication that we need to move through to do consensus problem solving.

First, we should generate ideas by discussion, using open questions and brainstorming creatively.

Second, we should record all the input somewhere so that everyone can see. Get everyone to understand others' ideas by reviewing what has been recorded. Then, work together to evaluate all the input. The challenge is to keep everyone focused on finding the solution and not wandering off on tangents.

Third, we should work out the best solution by summarizing, eliminating and narrowing. When we have narrowed down the possibilities, then we can quickly rank the solutions and prioritize what we think are the top solutions to the problem.

Following these three steps can lead to good, solid solutions to troubleshooting problems, once we get used to using the steps. What makes this approach work so well is that we get and consider everyone's ideas and don't overlook any solution. Sometimes there is a solution that no one has thought of before. By working together to pick the best solution, we will work better to implement the solution.

Relevant Tools and Equipment

- *National Electrical Code® 2008*
- Hand tools
- Multimeter

Final Code Review and Test Preparation (Student Manual) Lesson 435-6

👉 Code References

- Article 100
- Article 110
- Article 200
- Article 210
- Article 215
- Article 220
- Article 230
- Article 240
- Article 250
- Article 300
- Article 310
- Article 314
- Article 320
- Article 330
- Article 334
- Article 338
- Article 342
- Article 344
- Article 348
- Article 350
- Article 352
- Article 358
- Article 392
- Article 400
- Article 408
- Article 410
- Article 430
- Article 440
- 460-8
- Article 500
- Article 501
- Article 502
- Article 503
- Article 504
- Article 511
- Article 513
- Article 514
- Article 515
- Article 516
- Article 550
- Article 600
- Article 680
- Article 720
- Article 725
- Article 760
- Chapter 9 Table 4
- Chapter 9 Table 5
- Annex C

Homework

📖 1) Reading Assignment

- Review all your homework assignments for Lessons 404-43

A𝐴 2) Key Terms

Review all of the terms for Lessons 404-434.

Final Code Review and Test Preparation (Student Manual) Lesson 435-7

3) Practice Exercises

Answer the following referenced questions in the texts or other materials:
- None

Study Tips

▶ Here are a few hints to remember when you take tests like the Final Exam:

1) Be sure that you understand the instructions.

2) Be sure to read all the answers if it is multiple choice.

3) Don't struggle with a question and waste your time. Go on and come back to it later.

4) Make sure that you allow enough time to finish all the questions.

5) If you are unsure of an answer, put down what you think is the best answer and come back to it later. Something in the rest of the test may trigger the correct answer for you.

Semester Exam (Student Manual) Lesson 436-1

| *I E C* *PRIDE* *NATIONAL* | **Electrical Curriculum** **Year Four Student Manual** |

Lesson 436 – Semester Exam

To evaluate your progress throughout the semester, you will take a final exam of approximately 100 questions from the test banks for Lessons 419 through 434.

Glossary of Terms

GLOSSARY OF TERMS

IEC APPRENTICESHIP PROGRAM
FOURTH YEAR

ABDOMINAL THRUSTS—A technique used to dislodge a piece of food or other object from a person's airway.

ABSORPTION—Through the skin—One of the four ways that poison can enter the body.

ACTUATION—The process of putting into motion.

ADDRESSABILITY—Ability to direct data messages to the proper circuit, including the location of the device and its status, and to receive data back.

ADDRESSABLE CONTROL DEVICE—A device included in a fire alarm system that is used to individually control other functions.

ADDRESSABLE INITIATING DEVICE—A circuit device included in a fire alarm system that sets the system in motion.

AHJ—Authority having jurisdiction. Whoever is responsible for approving equipment installations and/or procedures.

AIDS—Acquired Immune Deficiency Syndrome.

ALARM NOTIFICATION APPLIANCE—The sound and/or light signaling part of a smoke or fire detector that will alert building occupants to the presence of fire or smoke.

ANODE—One of the three electrodes used as part of the normal operation of a silicon-controlled rectifier (SCR).

ANALOG—Referring to electrical signals that have continuously varying qualities of length and magnitude.

APPROVED—According to OSHA, the conductors and equipment required or permitted, by the authority enforcing Subpart S, Electrical of OSHA requirements. The authority is the Assistant Secretary of Labor for Occupational Safety and Health.

ARTERIES—Major blood vessels carrying blood from the heart and lungs to other parts of the body.

AUXILIARY TRIP RELAY—An output device attached with an initiating device. It is only activated when its associated initiating device is in the alarm state.

AUXILIARY FIRE ALARM SYSTEMS—Secondary systems that are in place in case the primary system should fail.

AVULSION—A cut in which a piece of skin or other tissue is partially or completely torn away.

BRAINSTORMING—A group process of quickly gathering possibilities about a topic without stopping to evaluate each one.

BRAKING—Any number of ways that are used to stop a motor more quickly than coasting.

BRUISE—Contusion —Injury that causes bleeding below the surface of the skin.

CAPILLARY—Minor blood vessel.

CATHODE—One of the three electrodes used in the normal operation of silicone-controlled rectifiers (SCR). A negatively charged electrode in an electrolytic cell, electron tube or storage battery.

CHEST COMPRESSION—Part of the CPR lifesaving method. The rhythmic application of pressure to the chest and heart in order to restore pulse and breathing.

CPR—Cardiopulmonary resuscitation—A combination of chest compressions and rescue breathing.

CURRENT SINKING OUTPUT—An output device that uses a NPN transistor as a switching element.

CURRENT SOURCING OUTPUT—An output device that uses a PNP transistor to detect the target.

CUTOFF REGION—The point where the transistor is switched off so that no current flows.

DATA I/O—Part of an input/output system of a PLC, providing complex information.

DIGITAL SIGNALING SYSTEMS—Alarm systems that contain one or more digital alarm communicator transmitters and one or more digital alarm communicator receivers.

DISCRETE I/O—Part of the input/output system of a PLC. Discrete parts manufacturing has individual parts made that can then be assembled by the use of a programmable controller (PLC).

DISLOCATION—A movement of a bone away from its normal position in a joint.

DUCT SMOKE DETECTOR—There is more than one type of detector that is installed by or within ducts that will signal if smoke is detected.

ECKO—Eddy Current Killed Oscillator principle that states that an oscillator works on a varying alternating magnetic field, even if a magnetic target is not present.

ELECTRICAL NOISE—Unwanted signals that enter through power supply lines and I/O devices.

Glossary-3

EMS—Emergency Medical Services System—An emergency response network of police, fire, and medical personnel.

FIBER OPTICS—A technology that uses transmission of light to connect electronic circuits by means of thin glass tubing or plastic fibers.

FOILS—Traces (definition in text).

FRACTURE—A complete break or crack in a bone.

GATES—The types of chips that are found in digital electronics. Devices that control electronic signals.

GENERAL ALARM— An alarm that alerts the building occupants to danger without specifying the exact cause or location. A fire alarm such as siren, horn, buzzer, visible annunciator and strobe light.

GOOD SAMARITAN LAWS—Laws adopted by some states that protect citizens, who administer aid to victims, from lawsuit.

GROUND FAULT DETECTOR—In a fire system, a ground-fault monitoring circuit indicates or identifies any installation conductors when they are grounded.

HAZCOM HAZARDOUS COMMUNICATION—Training or other methods used to inform workers of possible hazardous materials in their work area.

HEART ATTACK—This occurs when damage to the heart causes it to stop working effectively.

HEIMLICH MANEUVER—A lifesaving method that uses abdominal thrusts to dislodge foreign matter from the windpipe.

HIGH-LEG—In a three phase, four wire delta connection where the midpoint is grounded, one phase winding will have a higher voltage to ground and it must be an orange color conductor or the conductor must be tagged.

HIV—Human Immunodeficiency Virus—Thought to be the preliminary infection leading to AIDS.

HORN—Audible fire alarm that is used when a louder and/or more distinct signal is required.

HYBRID SOLID STATE RELAYS—Relays that are a combination of electromechanical and solid state technology that are used to solve unusual problems.

HYPERGLYCEMIA—An abnormally large concentration of sugar in the blood. Can cause diabetic coma.

HYPOGLYCEMIA—An abnormally low concentration of sugar in the blood. Can cause insulin reaction.

IEC—1) Independent Electrical Contractors, Inc. A national trade organization for independent electrical contractors. 2) Also a national standardization organization (International Electrotechnical Commission) that produces international standards for electrical and electronic equipment.

IMBALANCE—A lack of balance.

INCISION—A cut or laceration.

INDEPENDENT—Free of or not needing outside control.

INFECTION—Afflicted with germs or a virus.

INGESTION—By swallowing. One of the four ways that poison can enter the body.

INJECTION—By stings, bites, or hypodermic needle. One of the four ways that poison can enter the body.

INPUT-OUTPUT STATUS INDICATOR—It indicates the current status of input and output devices.

INTRINSICALLY SAFE CIRCUIT—An assembly of interconnected intrinsically safe apparatus, etc. that may be used in hazardous locations.

LACERATION—A tear, cut or incision in the skin.

LASER—A device that emits a beam of coherent visible radiation. A laser diode has an optical cavity that allows it to lase coherent light.

LOCAL FIRE ALARM SYSTEMS—Now called protective premises fire alarm systems. In NFPA 72, the definition of protected premises is "the physical location protected by the alarm system."

MARCH TIME CODED SYSTEM—A noncoded fire alarm system that produces an alarm signal at the rate of 120 pulses a minute.

MASTER CODED SYSTEM—A coded fire alarm system that gives the same common-coded signal from any location in the building.

MODULATED—Varied or able to be adjusted. In photoelectric controls, a light source that turns the control on and off at high speed.

MSDS—Material Safety Data Sheets—Used to provide necessary safety information regarding the use of chemicals.

MULTISPEED—Using a variety of speeds, such as in a motor that is required to run at different speeds.

OSHA—Occupational Safety and Health Administration—Federal Government organization responsible for producing and enforcing regulations regarding workplace safety and health.

PADS—Small round conductors to which component leads are soldered on a PC board.

PC BOARDS—Insulating material such as fiberglass, on which conducting paths are laminated.

PHASE LOSS—A situation that occurs when one (or more) fuse blows and the motor is single-phasing.

Glossary-5

PHASE REVERSAL—Can be caused by improper phase sequence and will result in damage to the motor and injury.

PHOTOELECTRIC-TYPE SMOKE DETECTOR—Any smoke detector that uses photoelectric light or a light source projected on a photosensitive device that activates an alarm when smoke to diffuse the light source.

PHOTO SENSOR—A photoreceiver. One of the two main components of a photoelectric switch.

PROBLEM SOLVING—In groups, a four part method of solving problems that consist of identifying the problem, determining its nature and cause, talking about the possible ways of solving the problem and finally, selecting the best solution.

PROPRIETARY SUPERVISING STATION FIRE ALARM SYSTEMS—Systems serving one or more properties which are adjacent or at a distance that are under one ownership.

REMOTE SUPERVISING STATION FIRE ALARM SYSTEMS—Systems that are connected to a municipal communications center or centers that relay an alarm. The two most common systems use "reverse-polarity" or differential current relays.

RESCUE BREATHING—A way of breathing air into a person to supply the oxygen he or she needs to survive.

SCAN—The process of going through all of the necessary steps to evaluate, execute and update an input/output program for a PLC.

SEIZURE—Loss of body control because of disruption in the brain. Can be caused by injury, fever, disease, chemical imbalance or infection.

SELECTIVE CODED SYSTEM—A fire alarm system that emits a alarm signal to identify specifically where a fire has broken out within a building. It notifies the proper fire-fighting personnel, also.

SERVICE DROP—Overhead cables that run from the power company's lines to the customer's service entrance conductors.

SERVICE LATERAL—Underground cables that run from the street main to the first point of connection to the service entrance conductors.

SIMULATION—Another term for role-playing: acting out a scenario or problem in order to come to possible conclusions.

SPLINT—Temporary reinforcement for a broken bone.

STRAIN—A stretching or tearing of muscles or tendons.

SUPPRESSION—The method used to dissipate arcing across opening contacts.

UNMODULATED—Unvarying or not able to be adjusted. In a photoelectric transmitter, the light beam is constantly transmitting.

VEIN—Blood vessel that returns deoxygenated blood to the heart and lungs.

VOICE FIRE ALARM SYSTEM—A fire alarm system that includes loud speakers and recorded sound or live voices.

VOLTAGE DROP DISSIPATION—In a switching component, the heat produced by a voltage drop will either be dissipated through the relay's case or by the use of a heat sink.

ZONED CODED SYSTEM—An audible alarm system that identifies the zone where a fire is detected as opposed to a specific area.

WORKSHEETS

Worksheet Lesson 401

1. List four key things to remember that will enhance your ability to speak effectively with others.

2. What is OSHA?

3. What is MSDS?

4. What are four tips you can use to memorize new material when studying?

5. What are three places where you might find safety requirements or ways to stay safe on the job?

6. What is IEC?

Worksheet Lesson 402

1. List the top five leading causes of death for people ages 0-44 years starting with the highest frequency.

2. In an emergency situation, what is the most important thing any person can do?

3. What are your four basic steps in an emergency situation?

4. What are four things that will help you identify an emergency?

5. When should you move a seriously injured person?

6. List the four steps when calling EMS.

7. When the airway of an unconscious person is blocked, you should get _____ in instead of getting the _____ out.

Worksheet Lesson 403

1. List four items that should be in a first aid kit.

2. List three things that you can do to prevent heat related illness.

3. What is often the first signal of heat related illness?

4. List four signals of late stage heat related illness.

5. What are three things you can do to try to prevent poisoning?

6. What are two symptoms of hypothermia?

7. What are two symptoms of frostbite?

Worksheet Lesson 404-1

1. AND gate

2. OR gate

3. NOR gate

4. NAND gate

5. Zenor diode

6. PIN photodiodes

7. Laser diode

8. 555 timer

9. DIAC

10. TRIAC

a. integrated circuit designed to output timing pulses.

b. a three-electrode AC conductor switch.

c. a diode similar to an LED but has an optical cavity.

d. a diode with large intrinsic region sandwiched between p-type and n-type regions.

e. a three-layer bidirectional device used primarily as a triggering device.

f. a device that provides a low output when either or both inputs are high.

g. a device with an output that is high when either or both its output is high.

h. acts as a voltage regulator.

i. a device that provides a low output when both inputs are high.

j. a device with an output that is high when only both of its inputs are high.

Worksheet Lesson 404-2

1. What can be used to connect pads?

2. A semiconductor is a material that has _____ in its outer orbit.

3. What is the process of adding electrons from other elements in the orbit of crystal's atoms?

4. A full-wave bridge rectifier may be made by using what?

5. In order to have proper polarity, what must happen with a diode?

6. A TRIAC function is similar to the action of _____ .
    a. SCRs
    b. LEPs
    c. PNPs
    d. IECs

7. In an amplifier, the relationship of the output signal to the input signal is known as what?

8. The Gantt chart gives a visual display of _____ associated with a project.

Worksheet Lesson 405

1. Reed relay
2. General purpose relay
3. Machine control
4. Solid state relay
5. Heat sink
6. Drop-out voltage
7. Insulation
8. Control current
9. Turn-off time
10. Switch-on voltage Drop

a. minimum current required to turn ON solid-state control circuit.

b. a device that forces cooling.

c. EMRs that include several sets of non-replaceable NO and NC contacts that are activated by a coil.

d. voltage drop across SSR when operating.

e. a fast-operating, single-pole single throw switch with NO contacts hermetically sealed in a glass envelope.

f. elapsed time between removal of control voltage and removal of voltage from load circuit.

g. an EMR that includes several sets of NO and NC replacement contacts that are activated by a coil.

h. amount of resistance measured across relay contacts in open position.

i. a relay that is low in cost, high reliability, and immense capability.

j. maximum voltage at which the relay is no longer energized.

Worksheet 406

Worksheet Lesson 406

1. What can operate on as little as five VDC?

2. What are throws?

3. Why must you take special precautions when using SSRs to control programmable controller inputs?

4. In an environment where arcless switching is required, what kind of relay should be used?

5. What is faster with a solid state relay than an electromechanical relay?

6. List two strengths and two weaknesses of the Gantt Chart.

Worksheet Lesson 407

1. What is the length of time it takes an object that is 4 inches wide to pass a ¼ inch diameter beam when the object is moving 1 ¼ inches per second?

2. What is the length of time it takes an object that is 6 ¼ inches wide to pass a ⅛ inch diameter beam when the objects is moving 2 ¼ inches per second?

3. What is the length of time it takes an object that is 3 ¾ inches wide to pass a 1/16 inch diameter beam when the object is moving 1 ¾ inches per second?

4. What is the length of time it takes an object 2 ½ inches wide to pass a ⅛ inch diameter beam when the object is moving 2 inches per second?

Worksheet Lesson 408

1. A programmable controller is designed for use where?

2. What does the processor section of a programmable controller do?

3. What are basic circuit logic functions used for?

4. Explain batch process control systems.

5. What kind of signals can a multiplexing system transmit?

6. What four elements make up a PERT chart?

Worksheet Lesson 409

1. _____ scanning is the first choice for edge guarding or positioning clear or translucent materials.

2. What does light-operated mean?

3. Explain the "skin effect" of ECKO sensors.

4. What kind of hall device is activated upon current flow through a conductor?

5. List two reasons why you should develop charts like the PERT chart.

6. List three advantages and three disadvantages of a PERT chart.

Worksheet Lesson 410

_____ 1. MSDS

_____ 2. Integrated circuit

_____ 3. Cutoff region

_____ 4. Rectifiers

_____ 5. Pole

_____ 6. Hybrid solid state relay

_____ 7. Discrete I/O

_____ 8. Interface

_____ 9. Dielectric

_____ 10. Current signaling output

_____ 11. Current sourcing output

_____ 12. Pareto Principle

_____ 13. ECKO

A. An insulator.

B. A particular point in a transistor where the current flow stops after the switch is turned off.

C. Also known as the 80-20 rule.

D. A circuit that is made up of different types of devices made from the same type of material.

E. Can control a switching procedure by using a NPN.

F. Brings in both electromechanical and solid state to solve unusual problems.

G. Must be completed on work place hazardous materials.

H. Alternating currents change to direct currents.

I. Using a PNP transistor to find a particular point.

J. An Eddy Current Kill Oscillator.

K. A supporting column that is circular.

L. A boundary that is shared.

M. A programmable controller that is able to manufacture parts.

Worksheet Lesson 411

1. What is the starting current of a DC motor with an armature resistance of .75 Ω that is connected to a 150 volt supply?

2. What is the starting current of a DC motor with an armature resistance of 2.25 Ω that is connected to a 180 volt supply?

3. What is the current during starting of a DC motor with an armature resistance of 1.25 Ω that is connected to a 100 volt supply and is generating 50 volts of EMF?

4. What is the current during starting of a DC motor with an armature resistance of .5 that is connected to 75 volt supply and is generating 40 volts of EMF?

5. What is the running current of a DC motor with an armature resistance of 1.25 that is connected to a 100 volt supply and is generating a 80 volt EMF?

6. What is the running current of a DC motor with an armature resistance of .5 Ω that is connected to a 75 volt supply and is generating a 60 volt EMF?

7. What is the synchronous speed of a motor having 8 poles that is connected to a 100 Hz power supply?

8. What is the synchronous speed of a motor having 6 poles that is connected to a 50 Hz power supply?

Worksheet Lesson 412

1. The formula for figuring braking torque is what?

2. What two things are proportional to horsepower?

3. How can the speed of a DC series motor be controlled?

4. A motor that must slow from high speed before being run at low speed has what kind of circuit logic?

5. What are two other names for friction brakes?

6. What are friction surfaces on friction brakes called?

7. What is the formula to determine the amount of work done?

8. What is the formula to determine power?

Worksheet Lesson 413

1. A motor with a synchronous speed of 1800 rpm and 2.8% slip factor has a full load speed of how many rpms?

2. Primary resistor starters provide _____ starting as the motor accelerates.

   a. fast
   b. rough
   c. extremely smooth
   d. none of the above

3. An autotransformer starter may provide more _____ on the load side than what is present on the line side.

   a. Current
   b. Voltage
   c. kvar
   d. KVA

4. To overcome the problem of unequal current division at normal operation of a dual voltage, delta connected motor, some part winding starters are furnished with a _____ -pole starting contactor and a _____ -pole running contactor.

   a. 6, 3
   b. 2, 4
   c. 3, 1
   d. 4, 2

5. The voltage across the windings in the wye configuration of a wye - delta starter is approximately what percent of line-to-line voltage?

6. List the five characteristics of a good task.

Worksheet Lesson 414

1. What does phase imbalance do to motors?

2. When should a ground test be performed on control transformers?

3. If you check a capacitor with an ohmmeter and observe the meter swinging to zero ohms, what does this indicate?

4. What five criteria should every goal meet?

5. List the four main reasons for a preventative maintenance program.

6. List the six items that should be included in a preventative maintenance program.

Worksheet Lesson 415

1. The formula for figuring breaking torque is what?

2. What is the formula to determine the amount of work done?

3. What is the formula to determine power?

4. Primary resistor starters provide _____ starting as the motor accelerates.

    a. fast
    b. rough
    c. extremely smooth
    d. none of the above

5. To overcome the problem of unequal current division at normal operation of a dual voltage, delta connected motor, some part winding starters are furnished with a _____ -pole starting contactor and a. _____ -pole running contactor.

    a. 6, 3
    b. 2, 4
    c. 3, 1
    d. 4, 2

6. List the four main reasons for a preventative maintenance program.

Worksheet 416

Worksheet Lesson 416

_____ 1. TELL

    a. Manager identifies the problem, consults with subordinates, then makes final decision.

_____ 2. SELL

    b. Manager defines the problem and its limitations, and then passes to the group (including self) the right to make the final decision.

_____ 3. CONSULT

    c. Manager identifies problem, considers alternative solutions, chooses one, then reports this decision for implementation.

_____ 4. JOIN

    d. Manager identifies problem, considers alternative solutions, chooses one by persuading subordinates to accept decision.

Copyright © 2008 by the Independent Electrical Contractors, Inc. All rights reserved.

Worksheet Lesson 417

_____ 1. Pole

_____ 2. Discrete I/O

_____ 3. Dielectric

_____ 4. Pareto Principle

_____ 5. ECKO

_____ 6. Hybrid solid state relay

_____ 7. Integrated circuit

_____ 8. Interface

_____ 9. Braking

_____ 10. Torque

_____ 11. Part-wind starting

_____ 12. Phase loss

_____ 13. Phase reversal

_____ 14. Goals

_____ 15. WBS

A. Represents work breakdown structure.

B. Brings in both electromechanical and solid state to solve unusual problems.

C. A supporting column that is circular.

D. In a motor, it is the turning effort.

E. Can lead to problems in electric motors because of an improper phase sequence.

F. An insulator.

G. A motor that is running on single-phase because one of the three phase was dropped.

H. Without coasting, a way to stop a motor.

I. A programmable controller that is able to manufacture parts.

J. Should be measurable and have a timeline for completion.

K. Also known as the 80-20 rule.

L. Has two or more circuits in the stator winding.

M. An Eddy Current Kill Oscillator.

N. A boundary that is shared.

O. A circuit that is made up of different types of devices made from the same type of material.

Copyright © 2008 by the Independent Electrical Contractors, Inc. All rights reserved.

# Worksheet Lesson 419

1. What is the difference between a temporary and a sustained power interruption?

2. Which is usually more destructive - voltage sags or voltage swells?

3. What is a harmonic?

4. How many cycles does a harmonic waveform at 180 Hz complete for each cycle of a fundamental frequency at 60 Hz?

5. What is the main difference between a linear and a nonlinear load?

6. Noise can enter a power distribution system from which sources?

7. What is the power factor?

8. What is a sign of possible overvoltage in an electrical system?

9. What is a solution for overvoltage in an electrical system?

10. What is a solution for noise in an electrical system?

Worksheet Lesson 420

1. Which Code section sets minimum lighting loads by occupancy?

2. Which conductor is specified as the conductor to be grounded for a single-phase, 3-wire AC premises wiring system?

3. Which Code section requires grounding for a 3-phase, 4-wire, wye connected, AC electrical system of 50 to 1,000 volts that supplies premises wiring and premises wiring systems?

4. Where a secondary tie is used to connect two transformers, what rating is required for the overcurrent device provided in the secondary connection of each transformer?

5. Which Code section requires sufficient diffusion and ventilation of gases from batteries to prevent the accumulation of an explosive mixture?

6. Which Code section defines Class 1, Class 2, and Class 3 circuits?

7. Conductors and cables of intrinsically safe circuits not in raceways or cable trays shall be separated at least how far from conductors and cables of any nonintrinsically safe circuits?

8. How should raceways, cable trays, and other wiring methods for intrinsically safe systems wiring be identified?

Worksheet Lesson 421

1. Internal connections of three-phase alternators are in a _____ configuration.

   a. Wye
   b. A and B
   c. Delta
   d. None of the above

2. A delta-connected transformer with one center tap normally is used to deliver _____, _____, and _____.

   a. 1PH-240V, 3PH-208V, 1PH-120V
   b. 1PH-120V, 1PH-240V, 3PH-208V
   c. 1PH-120V, 3PH-240V, 1PH-240V
   d. None of the above

3. Most substations normally incorporate a _____ to allow for line losses to the substation.

   a. Voltage regulator
   b. Current regulator
   c. Transformer output regulator
   d. None of the above

4. Article _____ of the *NEC®* states an assembly of one or more enclosed sections having a common power bus and principally containing motor control units is a

   a. Article 110, power panel
   b. Article 100, motor control center
   c. Article 90, power panel
   d. Article 120, motor control center

Worksheet Lesson 422

1. What is the minimum clearance required for switchgear supplied with 9010 volts installed opposite an insulated wall?

2. What is the minimum clearance required for switchgear supplied with 8900 volts installed opposite an insulated wall?

3. What is the minimum clearance for a motor control center supplied with 7260 volts installed opposite a concrete wall?

4. What is the minimum clearance for a motor control center supplied with 14,000 volts installed opposite a concrete wall?

5. What is the minimum clearance for equipment enclosures supplied with 7 kV installed opposite one another?

6. What is the minimum clearance for equipment enclosures supplied with 23,400 volts installed opposite one another?

Worksheet 423-1

Worksheet Lesson 423-1

1. A conductor's allowable ampacity is based upon which four determining factors?

2. Conductors routed through ambient temperatures greater than how many degrees Fahrenheit should be re-rated according to which section of the *NEC®*?

3. What should be done to compute continuous loads?

4. What causes overcurrents?

5. What is the "rule of thumb" used to determine when a neutral load is current carrying?

6. Insulated conductors of 6 AWG and smaller wire, when used as grounded (neutral) conductors, shall have what color insulation?

7. The ungrounded (phase) conductors shall be permitted to be identified with any color of insulation or tagging material except which colors?

8. What is the "rule of thumb" for determining the short term rating of a single conductor installed in a raceway or routed by itself?

Copyright © 2008 by the Independent Electrical Contractors, Inc. All rights reserved.

Worksheet Lesson 423-1 Cont.

9. What is the allowable ampacity of 11 current-carrying 6 AWG THW copper conductors routed through an ambient temperature of 95° F?

10. What size overcurrent protection device is required for 7-12 AWG THNN copper conductors that are run through an ambient temperature of 90° F?

Worksheet 423-2

Worksheet Lesson 423-2

1. Which Code section is about protection of conductors?

2. Which Code section is about protection of equipment?

3. Which Code section addresses remote control signaling and power-limited circuit conductor protection?

4. Which Code section addresses protection of fire alarm circuit conductors?

5. The overcurrent device of the branch circuit may protect the fixture wire for a 40-amp circuit if it is sized and selected per which Code section and uses what size AWG conductor?

6. Is a second stage of overcurrent protection required for extension cords that are 16 AWG and larger and connected to a 20-amp branch circuit?

7. Which section(s) of Code specify OCPD size for a known load?

8. Which section of Code gives the OCPD requirements for service-entrance conductors?

9. Which section of Code concerns sizing, loads, and operation of fuses?

10. Which section of Code concerns sizing, loads, and operation of circuit breakers?

Worksheet Lesson 423-3

1. What size time-delay fuse (min. size, next size, max. size) is required for a 30 HP, 230 volt, three-phase motor with a full-load rating of 60 amps?

2. What size time-delay fuse (min. size, next size, max. size) is required for a 15 HP, 230 volt, three-phase motor with a full-load rating of 42 amps?

3. What size time-delay fuse (min. size, next size, max. size) is required for a 25 HP, 230 volt, three-phase with a full-load rating of 52 amps?

Worksheet Lesson 423-4

1. What size nontime-delay fuse (min. size, next size, max. size) is required for a 45 HP, 230 volt, three-phase motor with a full-load rating of 58 amps?

2. What size nontime-delay fuse (min. size, next size, max. size) is required for a 20 HP, 230 volt, three-phase motor with a full-load rating of 40 amps?

3. What size nontime-delay fuse (min. size, next size, max. size) is required for a 60 HP, 230 volt, three-phase motor with a full-load rating of 74 amps?

Worksheet Lesson 423-5

1. What size circuit breaker (min. size, next size, max. size) is required for a 35 HP, 230 volt, three-phase motor with a full-load rating of 55 amps?

2. What size circuit breaker (min. size, next size, max. size) is required for a 20 HP. 230 volt, three-phase motor with a full load rating of 42 amps?

3. What size circuit breaker (min. size, next size, max. size) is required for a 50 HP, 230 volt, three-phase motor with a full-load rating of 78 amps?

Worksheet Lesson 423-6

1. What size instantaneous circuit breaker (min.. size, max. size) is required for a 40 HP, 230 volt, three-phase motor with a full-load rating of 65 amps?

2. What size instantaneous circuit breaker (min.. size, max. size) is required for 25 HP, 230 volt, three-phase motor with a full-load rating of 45 amps?

3. What size instantaneous circuit breaker (min. size, max. size) is required for 15 HP, 230 volt, three-phase motor with a full-load rating of 30 amps?

Worksheet Lesson 424

1. What is the ampacity of 6- # 12 THWN copper wire routed through an area of 140°F?

2. What is the ampacity of 7- # 6 THHW copper wire routed through an area of 100°F?

3. What is the ampacity of 12- # 10 THW copper wire routed through an area of 135°F?

4. What is the ampacity of 25- # 14 THHN copper wire routed through an area of 115°F?

Worksheet Lesson 425-1

1. What is the count for 15 spliced conductors passing through an octagon box?

2. What is counted toward the fill space of two boxes where there are 2-12 AWG luminaire (fixture) wires without the use of a conductor and 3-16 AWG luminaire (fixture) wires plus an equipment grounding conductor?

3. How should equipment grounding conductors passing through or spliced together in a box be counted?

4. Which section of Code specifies trade sizes, minimum volume, and maximum number of conductors for metal boxes?

5. Since there are no tables available to calculate the number of conductors permitted in junction boxes, how should an installer calculate the size of a junction box for conductors that are the same size?

6. Since there are no tables available to calculate the number of conductors permitted in junction boxes, how should an installer calculate the size of a junction box for conductors that are different sizes?

7. What section of Code should be used to select conduit for conductors that have the same type of insulation and that are the same size?

8. How should an installer select conduit for different size conductors?

Worksheet Lesson 425-1 Cont.

9. Which section of Code should be used to size nipples?

10. Which section of Code addresses cable tray selection, installation, and grounding and bonding?

Worksheet Lesson 425-2

1. Which section of the *NEC®* concerns the mounting and support of electrical boxes and junction boxes?

2. Which section(s) of the *NEC®* concern installation of multiconductor cables, such as armored cable (AC), metal-clad cable (MC), nonmetallic sheathed cable (romex), and service entrance cable?

3. Intermediate metal conduit (IMC) and rigid metal conduit (RMC), along with couplings, elbows, and fittings, shall be permitted to be installed where?

4. With the exception of some applications for trade size 3/8 (12), what is the permitted size range for flexible metal conduit?

5. What are the support requirements for electrical metal tubing?

6. Are approved grounding electrodes required for temporary construction poles used to mount meters and panel boards?

7. What section of Code gives the rules for temporary wiring?

8. Cable systems are only installed in floor areas under what conditions?

Worksheet Lesson 425-2 Cont.

9. Which section of Code gives the requirements for boxes and enclosures supported by framing members?

10. Is wiring run through the center of ceiling rafters considered supported?

Worksheet Lesson 426

1. What size octagon box is required to support a lighting fixture supplied with two #12 with ground non-metallic sheathed cables with a fixture stud, hickey, and two #16 fixture wires?

2. What size octagon box is required to support a lighting fixture with two #14-2 with ground non-metallic sheathed cables with a fixture stud, two cable clamps, two pigtails and two #16 fixture wires?

3. What size octagon box is required with cable clamps required for four #12-2 with ground romex cables that are spliced in the box between the panelboard and the loads?

4. What size square box is required for two #14-2 with ground connecting to a receptacle and four #12-2 with ground romex cables that pass through the box?

5. What size square box is required where two #14 and six #12 romex cables with ground are spliced and routed to loads in the building?

Worksheet Lesson 427

1. How many cycles does a harmonic waveform at 180 Hz complete for each cycle of a fundamental frequency at 60 Hz?

2. What is a solution for overvoltage in an electrical system?

3. Which conductor is specified as the conductor to be grounded for a single-phase, 3-wire AC premises wiring system?

4. Which Code section requires grounding for a 3-phase, 4-wire, wye connected, AC electrical system of 50 to 1,000 volts that supplies premises wiring and premises wiring systems?

5. A delta-connected transformer with one center tap normally is used to deliver _____, _____, and _____.

6. Article _____ of the *NEC*® states an assembly of one or more enclosed sections having a common power bus and principally containing motor control units is a _____.

7. What is the minimum clearance required for switchgear supplied with 8900 volts and installed opposite an insulated wall?

8. A conductor's allowable ampacity is based upon which four determining factors?

Worksheet Lesson 427 Cont.

9. Since there are no tables available to calculate the number of conductors permitted in junction boxes, how should an installer calculate the size of a junction box for conductors that are the same size?

10. How should equipment grounding conductors passing through or spliced together in a box be counted?

Worksheet Lesson 428

1. What are the only colors permitted for the wire connected to the green equipment grounding conductor terminal?

2. What section of Code gives the rules for replacement of receptacles?

3. What are the requirements for the supply of all laundry receptacle outlets?

4. Wall receptacle outlets shall be installed so that no point on a wall is further than what distance from a receptacle outlet?

5. Receptacle outlets shall be located how far from the inside wall of a spa or hot tub to ensure the safety of the user?

6. In commercial and industrial locations how shall the number of outlets on a general-purpose branch circuit be computed?

7. In residential dwelling units, how is the number of lighting outlets connected to a 15- or 20-amp general-purpose branch circuit determined?

8. How many lighting outlets are required for bathrooms?

9. Where should the lighting outlet be installed in utility rooms?

10. Which section of Code gives the requirements for switches and circuit breakers in wet locations?

Worksheet Lesson 429

1. What are the three types of currents and associated sections of Code that must be found before designing and selecting elements for circuits supplying power to motors?

2. Branch circuit conductors supplying a single motor shall have an ampacity not less than what percentage of the motor full-load current rating in amps?

3. The selection of conductors for wye start and delta run motors must be based on what percentage of the motor's full-load current, in amps, times 125% for continuous use?

4. What section(s) of Code are involved in sizing a motor branch-circuit overcurrent protective device?

5. What is the locked-current rating for a three-phase, 480 volt, 60 horsepower motor with a Code letter "G" marked on the nameplate of the motor?

6. What section of Code specifies the sizes and types of controllers required to start and stop motors?

7. The disconnecting means for motor circuits shall have an ampere rating of at least what percent of the full-load current rating of the motor per what section of Code?

8. How is the OCPD sized for two or more compressor motors?

9. How is the disconnecting means sized for compressor motors?

10. What Code section specifies methods for grounding compressor motors?

Worksheet Lesson 430

1. Which section of Code defines "Dust-Ignitionproof"?

2. Which of the following is a Class I, Division 1 location?

    (a) A dyeing plant using flammable liquids
    (b) The interior of an exhaust duct used to vent ignitable concentrations of gases or vapors
    (c) The interior of a refrigerator where volatile flammable liquids are stored
    (d) All of the above
    (e) None of the above

3. What is the classification for a location where combustible dust is not normally in the air in quantities sufficient to produce explosive or ignitable mixtures, but combustible dust may be in suspension in the air as a result of handling or processing equipment?

4. A flammable gas such as propane with a MESG value greater than 0.75 mm or a minimum igniting current ratio (MIC ratio) greater than 0.80 is classified in which material group?

5. Class II, Group G materials include what kind of hazardous materials?

6. Hermetically sealed equipment is an allowable type of protection for which hazardous locations?

7. What are the two factors taken into consideration to identify (approve) equipment for use in a hazardous location?

8. Which section of Code specifies wiring requirements for Class I, Division 1 locations?

9. Which section of Code specifies sealing methods for Class II locations?

10. Which section of Code specifies the separation of intrinsically safe conductors?

Worksheet Lesson 431

1. Which section(s) of Code specify when a commercial garage, repair, and storage facility are classified as a hazardous location?

2. Which section of Code specifically defines a Class I, Division 1 location?

3. Is a raceway in a masonry wall or buried beneath a floor in a garage permitted to have connections leading into Class I, Division 1 or 2 locations if the raceway is considered to be in a non-hazardous location?

4. If it is not in conduit or tubing, fixed wiring above Class I locations should be of which type?

5. All 125-volt, single-phase, 15- and 20-ampere receptacles installed in areas where electrical diagnostic equipment, electrical hand tools, or portable lighting equipment are used must have what feature?

6. Can aircraft battery chargers and their control equipment be operated in Class I locations as defined in *NEC®* 513.3?

7. For a remote outdoor gasoline pump, the Class I, Division 2 hazardous area extends outward from the edge of the gasoline dispenser to all directions for a minimum distance of how many feet?

8. A lighting fixture for use in a Class II, Division 1 area of a grain elevator should use what kind of support structure?

9. Is a light mounted 4 feet above the ceiling of a ventilated spray booth required to be listed for use in a Class I, Division 2 hazardous location?

10. Does a service area in a garage where vehicles are driven meet Code requirements if it uses standard fluorescent industrial fixtures with exposed lamps mounted at 12.5 feet above the floor?

Copyright © 2008 by the Independent Electrical Contractors, Inc. All rights reserved.

Worksheet Lesson 432

1. *NEC*® Article 600 covers what kinds of lighting?

2. Outlets for outline lighting systems must be supplied by what kind of circuit?

3. What are the two options for the disconnecting means for outline lighting systems?

4. Which section of Code concerns grounding outline lighting systems?

5. What size and type of bonding conductors are required for outline lighting systems?

6. In wet or damp locations, what do portable or mobile signs need to protect personnel?

7. What is the longest power supply cord permitted for portable or mobile signs in dry locations?

8. How much working space shall be provided for electric signs at each ballast, transformer, and electric power supply or its enclosure where not installed in a sign?

Worksheet Lesson 433-1

1. What is the VA and amps for a residential unit with the following loads?
   - 2800 sq. ft. living area
   - 3 small appliance circuits
   - 1 laundry circuit

   120 V, SINGLE-PHASE LOADS
   - 1,200 VA disposal
   - 1,300 VA compactor
   - 1,800 VA dishwasher
   - 2,200 VA microwave

   240 V, SINGLE-PHASE LOADS
   - 6,000 VA dryer
   - 11,000 VA A/C
   - 22,000 VA heating unit
   - 8,000 VA water heater
   - 9,000 VA oven
   - 9,000 VA cook top
   - 850 VA blower motor
   - 1,200 VA pool pump

2. What is the VA and amps for a residential unit with the following loads:
   - 1675 sq. Ft. of living area
   - 2 small appliance circuits
   - 1 laundry circuit

   120 V, SINGLE-PHASE LOADS
   - 2,000 VA water pump
   - 1,200 VA disposal
   - 1,200 VA compactor
   - 1,400 dishwasher

   240 V, SINGLE PHASE LOADS
   - 5,000 VA dryer
   - 9,000 VA A/C
   - 16,000 VA heating unit
   - 7,000 VA water heater
   - 10,000 VA oven
   - 8,000 VA cook top
   - 600 VA blower motor

Worksheet Lesson 433-2

1. What is the load in VA and amps to compute and size the elements for a 120/240 volt, single-phase service supplying a 35,000 sq. ft store with 15,000 sq. ft. of warehouse space?

   Given loads: (120 volt single-phase loads)
   - 50 linear feet of show room window (noncontinuous operation)
   - 40 ft. of track lighting
   - 2,000 VA sign lighting (continuous operation)
   - 50 receptacles (noncontinuous operation)
   - 30 receptacles (continuous operation)
   - 20 ft. multioutlet assembly (heavy duty)

   Given loads: (240 volt, single-phase loads)
   - 5,000 freezer
   - 4,000 VA ice cream boxes
   - I HP exhaust fan
   - 12,000 VA water heater
   - 10,000 VA walk-in cooler
   - 50,000 VA heating unit
   - 34,000 VA A/C unit

Worksheet Lesson 435 Cont.

9. If it is not in conduit or tubing, fixed wiring above Class I locations should be of which type?

10. What are the two factors taken into consideration to identify (approve) equipment for use in a hazardous location?

Worksheet 435

Worksheet Lesson 435

1. Wall receptacle outlets shall be installed so that no point on a wall is further than what distance from a receptacle outlet?

2. In commercial and industrial locations how shall the number of outlets on a general-purpose branch circuit be computed?

3. In residential dwelling units, how is the number of lighting outlets connected to a 15- or 20-amp general-purpose branch circuit determined?

4. What are the three types of currents and associated sections of Code that must be found before designing and selecting elements for circuits supplying power to motors?

5. Branch circuit conductors supplying a single motor shall have an ampacity not less than what percentage of the motor full-load current rating in amps?

6. The selection of conductors for wye start and delta run motors must be based on what percentage of the motor's full-load current, in amps, times 125% for continuous use?

7. The disconnecting means for motor circuits shall have an ampere rating of at least what percent of the full-load current rating of the motor per what section of Code?

8. What is the classification for a location where combustible dust is not normally in the air in quantities sufficient to produce explosive or ignitable mixtures, but combustible dust may be in suspension in the air as a result of handling or processing equipment?

Worksheet 433-3

Worksheet Lesson 433-3

1. In VA and amps, compute and size the elements for a 277/480 volt, three-phase, four-wire service supplying a 40,000 sq. ft. classroom area, 6,000 sq. ft. auditorium area, and a 1,500 sq. ft. assembly hall. (General lighting load supplied to fixtures is 277 volts.)

   Given Loads: (120 volt single-phase loads)
   - 210 receptacles (noncontinuous duty)
   - 70 receptacles (continuous duty)
   - 150 ft. multioutlet assembly (heavy duty)

   Given loads: (Single-phase and three-phase loads)
   - 5-1HP hood fans; 208 V, single-phase
   - 4 3/4 HP grill vent fans, 208 V, single-phase
   - 1 8-3A HP exhaust fans, 480 V, three-phase

   Given loads: (Cooking equipment)
   - 4-1kw toasters, 120 V, single-phase
   - 5-2kw refrigerators, 120 V, single-phase
   - 4-2kw freezers, 120 V, single-phase
   - 5-11kw ranges, 208 V, single-phase
   - 4-10kw ovens, 208 V, single-phase
   - 5-3.5kw fryers, 208 V, single-phase

   All loads are continuous

Worksheet Lesson 434

1. Which Code section sets minimum lighting loads by occupancy?

2. Which Code section defines Class 1, Class 2, and Class 3 circuits?

3. Which Code section requires grounding for a 3-phase, 4-wire, wye connected, AC electrical system of 50 to 1,000 volts that supplies premises wiring and premises wiring systems?

4. Which section of Code gives the requirements for switches and circuit breakers in wet locations?

5. What section(s) of Code are involved in sizing a motor branch-circuit overcurrent protective device?

6. What Code section specifies methods for grounding compressor motors?

7. Which section of Code specifies wiring requirements for Class I, Division 1 locations?

8. Which section of Code specifies sealing methods for Class II locations?

9. Which section of Code specifies the separation of intrinsically safe conductors?

10. Which section of Code specifically defines a Class I, Division 1 location?

Worksheet Lesson 435

1. Wall receptacle outlets shall be installed so that no point on a wall is further than what distance from a receptacle outlet?

2. In commercial and industrial locations how shall the number of outlets on a general-purpose branch circuit be computed?

3. In residential dwelling units, how is the number of lighting outlets connected to a 15- or 20-amp general-purpose branch circuit determined?

4. What are the three types of currents and associated sections of Code that must be found before designing and selecting elements for circuits supplying power to motors?

5. Branch circuit conductors supplying a single motor shall have an ampacity not less than what percentage of the motor full-load current rating in amps?

6. The selection of conductors for wye start and delta run motors must be based on what percentage of the motor's full-load current, in amps, times 125% for continuous use?

7. The disconnecting means for motor circuits shall have an ampere rating of at least what percent of the full-load current rating of the motor per what section of Code?

8. What is the classification for a location where combustible dust is not normally in the air in quantities sufficient to produce explosive or ignitable mixtures, but combustible dust may be in suspension in the air as a result of handling or processing equipment?

Worksheet Lesson 435 Cont.

9. If it is not in conduit or tubing, fixed wiring above Class I locations should be of which type?

10. What are the two factors taken into consideration to identify (approve) equipment for use in a hazardous location?